Teachers! Prepare Your Students for the
Mathematics for SAT*I

Teachers! Prepare Your Students for the Mathematics for SAT*I

Methods and Problem-Solving Strategies

Alfred S. Posamentier
Stephen Krulik

CORWIN PRESS, INC.
A Sage Publications Company
Thousand Oaks, California

Copyright © 1996 by Corwin Press, Inc.

All rights reserved. No part of this book may be reproduced or utilized in any form or by any means, electronic or mechanical, including photocopying, recording, or by any information storage and retrieval system, without permission in writing from the publisher.

For information address:

Corwin Press, Inc.
A Sage Publications Company
2455 Teller Road
Thousand Oaks, California 91320
e-mail: order@corwin.sagepub.com

SAGE Publications Ltd.
6 Bonhill Street
London EC2A 4PU
United Kingdom

SAGE Publications India Pvt. Ltd.
M-32 Market
Greater Kailash I
New Delhi 110 048 India

Library of Congress Cataloging-in-Publication Data

Posamentier, Alfred S.
    Teachers! prepare your students for the mathematics for SAT I : methods and problem-solving strategies / Alfred S. Posamentier, Stephen Krulik.
       p.   cm.
    ISBN 0-8039-6481-1 (alk. paper) — ISBN 0-8039-6416-1 (pbk : alk. paper)
    1. Mathematics — Examinations, questions, etc. 2. Scholastic Aptitude Test — Study guides. I. Krulik, Stephen. II. Title
QA43.P645 1996
510'.76 — dc20   96-5312

This book is printed on acid-free paper.

96 97 98 99 00 10 9 8 7 6 5 4 3 2 1

Corwin Press Production Editor: S. Marlene Head

CONTENTS

Preface vii
About the Authors viii
Introduction 1
  I. About the *SAT I* 4
    • Format of the *SAT I* 4
    • The Mathematics Section of the *SAT I* 5
    • Calculator Use on the *SAT I—Reasoning Test* 6
    • Student-Produced Response Questions 7
  II. Advising Students on Test Taking 8
    • Preparation 8
    • Test-Taking Procedures 8
    • To Guess or Not to Guess 9
  III. Nonroutine Mathematics Topics to Strengthen Students' Problem-Solving Abilities . 11
    • Using Proportional Relationships to Solve Problems 12
    • Simplifying the Procedure to Solve an Equation 13
    • Comparing Areas of Similar Figures 13
    • Work Problems 14
    • Averaging Rates 15
    • Finding Areas of "Nameless" Geometric Figures 17
    • Likelihood Problems 18
    • Successive Discounts 19
    • Clock Problems 20
    • Sequence Problems 22
    • Counting Problems 24
    • Comparing Fractions 25
  IV. Specific Problem-Solving Strategies 27
    Strategy 1: Working Backwards 27
    Strategy 2: Finding a Pattern 30
    Strategy 3: Adopting a Different Point of View 38
    Strategy 4: Solve a Simpler, Analogous Problem 48
    Strategy 5: Using Extreme Cases 54
    Strategy 6: Visual Representation 57

    Strategy 7: Intelligent Guessing and Testing 61
    Strategy 8: Accounting for All Possibilities 67
    Strategy 9: Organizing Data 72
    Strategy 10: Deductive Reasoning 78
Appendix: Review of Mathematical Principles 85
    • Rules for Divisibility 85
    • Order of Operations 87
    • A Brief Review of Fractions, Decimal Fractions, and Percents 88
    • Percentage Problems 92
    • A Brief Review of Some Essentials of Algebra 93
    • A Brief Review of Some Essentials of Probability 107
    • A Brief Review of Some Essentials of Geometry 109

PREFACE

For years the authors of this book have been deeply engaged in the promotion of problem solving as a skill to be introduced to students as an end in itself rather than as a means to an end (getting the answer to the problem). Both authors have for decades written material designed to prepare students for the SAT. Both authors have worked with students preparing for the SAT in regular classroom instruction and in specially designed SAT preparation courses. Perhaps more importantly, both authors have been working with both preservice and inservice mathematics teachers, where a significant emphasis was placed on the teaching of problem-solving skills. This book is the product of experiences gained during these many years of helping mathematics teachers acquire ways to teach the process of problem solving.

The present book discusses problem-solving methods and test-taking considerations, with which mathematics teachers should be familiar. Some of the illustrative problems presented are a bit beyond the scope of the *SAT I*; however, the points illustrated by these problems are better made with these problems. Teachers must, under any circumstances, tailor their instruction to their classes. Our illustrative problems are intended only to clarify the problem-solving methods presented, and not to be the exclusive examples of a given procedure.

We are aware that the traditional school curriculum, being topic driven, does not always provide a problem-solving emphasis. We have reviewed the curriculum to highlight currently taught concepts that can be particularly useful tools in solving mathematical problems. These concepts, such as proportionality, are taught in the context of specific topics and often are not applied beyond their classroom application. Yet, when seen as problem-solving tools, they can be very useful. We have also provided a very comprehensive overview of the entire mathematics material required for the *SAT I*. Both of these sections will be particularly useful to provide students with a refresher look back at previously taught material. These sections will also offer a concise way for the teacher to help students who have forgotten some of their earlier mathematics.

Throughout the book the discussions show the pitfalls that might be present and light the way to logical solutions. With this book as a guide, we believe that teachers ought to be well prepared to offer a meaningful problem-solving course in preparation for the *SAT I*, and not merely a continuous review of previous test items.

The authors wish to express their gratitude to Jacob Cohen and Ira Ewen, mathematics educators who read the manuscript and offered meticulous comments and suggestions. The artwork was carefully done by Roland Franzke, and for that he has our thanks. In particular, we wish to thank our editor, Joy E. Runyon, for her extraordinary dedication to the project.

Alfred S. Posamentier
Stephen Krulik

February 1996

About the Authors

ALFRED S. POSAMENTIER is Professor of Mathematics Education and Deputy Dean of the School of Education of the City College of the City University of New York. He is the author and coauthor of numerous mathematics books for teachers and secondary school students. As a guest lecturer, he favors topics regarding aspects of mathematics problem solving and the introduction of uncommon topics into the secondary school realm for the purpose of enriching the mathematics experience of those students. The development of this book reflects these penchants.

After completing his A.B. degree in mathematics at Hunter College of the City University of New York, he taught mathematics for six years at Theodore Roosevelt High School in the Bronx (New York), where he focused his attention on improving the students' problem-solving skills. He also developed the school's first mathematics teams (both at the junior and senior level) and established a special class that had as its primary focus methods of mathematics problem solving.

Immediately upon joining the faculty of the City College (after having received his masters' degree there), he began to develop extremely popular inservice and preservice courses for secondary-school mathematics teachers on the topic of mathematics problem solving.

Dr. Posamentier received his Ph.D. from Fordham University (New York) in mathematics education and has since extended his reputation to Europe. He is an Honorary Fellow at the South Bank University (London). He has been visiting professor at several British, German, and Austrian universities, most recently at the University of Vienna and at the Technical University of Vienna; at the former he was a Fulbright Professor.

In recognition of his outstanding teaching, Dr. Posamentier was recently named Educator of the Year by the City College Alumni Association and had a day named in his honor by the City Council President of New York City. More recently, he was awarded the National Medal of Honor from the Austrian government.

STEPHEN KRULIK is Professor and Coordinator of Mathematics Education at Temple University in Philadelphia, where he is responsible for the undergraduate and graduate preparation of mathematics teachers for grades K–12, as well as the inservice alertness training of mathematics teachers at the graduate level. He teaches a wide variety of courses, including the History of Mathematics, Methods of Teaching Mathematics, and the Teaching of Problem Solving.

Dr. Krulik received his B.A. in mathematics from Brooklyn College of the City University of New York and his M.A. and Ed.D. from Columbia University's Teachers College. Before coming to Temple University, he taught mathematics in the New York City public schools for 15 years. During this time, he created and implemented several courses designed to prepare students for the SAT examination and stressing the art of problem solving.

Nationally, Dr. Krulik has served as a member of the committee responsible for preparing the *Professional Standards for Teaching Mathematics* of the National Council of Teachers of Mathematics. He was also the editor of the NCTM's 1980 yearbook, *Problem Solving in School Mathematics*. Regionally, he served as president of the Association of Mathematics Teachers of New Jersey and was a member of the editorial team that produced their 1993 publication, *The New Jersey Calculator Handbook*.

His major areas of interest are the teaching of problem solving and reasoning, as well as comprehensive assessment in mathematics. He is the author and coauthor of more than 15 books for teachers of mathematics, including the forthcoming *The New Sourcebook for Teaching Reasoning and Problem Solving in the Junior and Senior High Schools*. He is also the senior problem solving author for a major basal textbook series. Dr. Krulik is a frequent contributor to the professional journals in mathematics education. He has served as a consultant to, and has conducted many workshops for, school districts throughout the United States and Canada. He is in demand as a speaker at both national and international professional meetings, where his major focus is on problem solving in school mathematics.

Introduction

In the early days of the Scholastic Aptitude Test, preparing students for the exam was frowned upon by the College Entrance Examination Board, the developers of the SAT. Before 1974, the Board claimed that the SAT was an examination to test *aptitude* (as the name suggests), a characteristic that the student could not improve by preparing in the relatively short time before the test; the Board claimed that this characteristic was developed over long periods of time as part of the normal mathematics instruction. In the mid-1970s, this notion was dispelled as a result of various research studies that showed (at least for the mathematics section) that practice prior to the test did have its advantages.

It was also clear that the student enrolled in a mathematics course during the semester in which he or she was to take the SAT would do significantly better than if that student were not enrolled in some mathematics course. Understandably, parents want to maximize their child's score on this test. For decades, privately run coaching courses have been very popular with students preparing for the SAT. With an increasing number of students going on to college, these courses became inadequate to meet the volume of need. Furthermore, parents who could not afford the cost for such coaching courses were at a disadvantage in their effort to provide their child with the broadest college selection afforded by good scores on the SAT. Consequently, high schools began to offer coaching courses for the SAT.

Some high schools thought that offering such a course was tantamount to admitting that their regular courses were inadequate. They then publicized that SAT preparation would take place in the regular mathematics courses, integrated into the coursework throughout the semester. Today, in addition to an increasing assortment of private coaching courses, there is a sharp increase of *SAT I* preparation done in the high schools. These are offered as one-semester courses, one-year courses, or six- or eight-week courses; sometimes they are even integrated into the regular mathematics offerings.

Just as some *SAT I* coaching or preparation courses are good and some are poor, there are effective and ineffective ways to integrate the *SAT I* preparation work into the syllabus of the regular courses. The purpose of this book is to propose various effective practices that should lead to improved performance on the mathematics section of the *SAT I*.

The key to a successful *SAT I* preparation course is the extent to which it meets the individual student's needs and the extent to which it focuses on true mathematics problem solving. In the former case, one has to assume that the classes are small enough for the teacher (hopefully an experienced one) to be able to meet the individual student's needs. There is little that can be done in the form of curriculum advice that can assist a teacher in this concern; the teacher should be aware of general methods of teaching that strive for this end. Books describing methods of teaching mathematics, such as *Teaching Secondary School Mathematics, Techniques and Enrichment Units* by Posamentier and Stepelman (4th edition, Merrill/Prentice Hall, 1995), will provide a host of methods for individualizing instruction or meeting individual needs in the more traditional setting.

In this book we focus on the curricular issues: what should be taught, how it should be presented, and in what setting it is most effective. It is well known that the *SAT I* is a timed test where accuracy and speed are essential for success. Accuracy is a habit we must hope the student will achieve through training and practice, and speed is a function of the tools with which the student attacks the problem. Just as a mechanic can do a better and faster job with power tools, so can a student better solve problems with an arsenal of more powerful problem-solving tools. This book is dedicated to achieving that end.

Consider a mechanic about to embark on a construction project. Before her is a wall full of all kinds of tools, neatly arranged so that they are all easily accessible. As she plans her work, the mechanic must select the tools she will use. If she is familiar with the all the tools on her wall, then she will select those that will most efficiently do the job. If some of the tools are unfamiliar, then the mechanic is somewhat handicapped, for she will not necessarily know the most efficient method for doing the work. To be successful in her business, she must efficiently—that is, quickly *and* correctly—do the work. The more familiar she is with these tools and the more adept she is at using them, the more successful she will be.

So, too, the person attempting to solve a mathematics problem has "tools" before him, in the form of problem-solving strategies. Some problems can be solved with any one of several strategies, whereas others clearly require one primary strategy. The more strategies students have at their disposal and the better able they are to discern which is the most appropriate in each situation, the better their chances of solving a problem in an efficient manner.

In this book we give teachers a host of problem-solving strategies to present to students. The strategies must be presented in a motivational fashion, usually done by showing how a given strategy makes a seemingly difficult or complex problem very simple to solve. Presenting lots of applications for these methods will help familiarize students with each strategy and produce lasting growth in students' problem-solving skills. Teachers should introduce them in categories, but eventually these methods should be presented in an integrated way so that students can practice selecting the best strategy for a particular problem.

Just as the mechanic refers to her tools and remembers them by their names, we, too, have titled the various problem-solving strategies. Referring to a strategy by name makes it easier to remember the strategy and to call on it when needed.

A good *SAT I* preparation course must do more than merely review test items in the hope that "practice makes perfect." There must be an organized presentation of material that gets at the heart of heuristics and at the same time provides students with the much-needed review and guidance to good problem-solving practices (sometimes referred to as "helpful hints").

This book is designed to focus on the aspects of what ought to be a wholesome preparation course for the *SAT I*. Such a preparation would include elements from the various components of this book:

- An overview of the *SAT I*, including a description of its format, its content, and the use of a calculator on the test
- A selective review of the mathematics taught through elementary algebra and geometry, with particular attention to problem solving
- Less well known mathematics "facts" and problem-solving short cuts, which can increase a student's arsenal of problem-solving tools
- A discussion on advising students on strategies for taking the *SAT I*, including the issue of when and how to guess on items unfamiliar to the student by considering the way the test is scored
- A detailed presentation of the following ten specific problem-solving strategies:
  1. Working backwards
  2. Finding a pattern
  3. Adopting a different point of view

4. Solving a simpler analogous problem
5. Using extreme cases
6. Visual representation
7. Intelligent guessing and testing
8. Accounting for all possibilities
9. Organizing data
10. Deductive reasoning

There is a particular philosophy with which this book is presented. First, the illustrations have been selected to demonstrate in dramatic form the power of the procedure presented. Second, the problems both in this book and the companion student book are slightly more difficult than the items on the *SAT I*. This is done in the same spirit with which basketball players run in practice sessions with weights on their ankles—so that by being trained under more strenuous conditions than those found in a regular game, they will perform better under the relatively less difficult game conditions.

I. ABOUT THE *SAT I*

The *SAT I–Reasoning Test* examines the student's potential for college work. It is clearly not the only indicator of potential. Though high school grades are probably the best predictor of how well a student will do in college, admissions officers also rely on the yardstick of standardized tests (such as the *SAT I*) when making admission decisions, because such tests compare students from many different schools and regions. Results of the *SAT I*, expressed as a score within a range of 200 to 800 and as a percentile rank, allow comparison with other college-bound students.

It would be impractical in a three-hour exam to test every subject area covered by high school courses. Rather, the *SAT I* is written in a way that supposedly tests verbal and mathematical reasoning ability to apply knowledge gained from these courses. The tests of verbal and mathematical reasoning are designed to reflect how developed the students' reading comprehension and problem-solving skills are and, consequently, how successful they will be in college.

FORMAT OF THE *SAT I*

There are three test sections of verbal reasoning and three of mathematical reasoning on the *SAT I*, separated into two main sections as follows:

- Two 30-minute sections and one 15-minute section of **verbal reasoning ability** (with an emphasis on critical-reading questions), comprising 78 questions in all.
- Two 30-minute sections and one 15-minute section on **mathematical reasoning ability**, comprising 60 questions in all. The questions in the latter section are of three types:

    35 multiple-choice questions, each having five possible choices
    15 quantitative comparison questions, multiple choice, each having four possible choices
    10 student-produced response questions, answered on a grid by blackening spaces with the solution determined by the test taker

There is also a section on either verbal or mathematical skills that is not counted in the examinee's score; it is used to ensure that *SAT I* scores remain consistent from test to test, and it "tries out" questions that may appear on future tests. The scores from the verbal and mathematics sections of the test are reported separately, so one score does not affect the other.

A final point about the format of the *SAT I*: Because it is a timed test, it is essential that students read directions quickly. It is probably wise to have students "memorize" the directions from their review sessions with the practice tests. (N.B.: The instructions in the accompanying students' book, *Students! Get Ready for the Mathematics for SAT I: Problem-Solving Strategies and Practice Tests*, have the same directions found on the actual *SAT I*.) The time saved on the test by not having to read the directions is a significant factor in taking the test. Students should judge rapidly which questions they can answer easily and which will require more time. Naturally, they should first answer the questions whose answers are more obvious, then proceed to the more difficult questions. It is very frustrating to have questions left unread when time runs out. (See "To Guess or Not to Guess" on p. 9.)

Students should obtain from the College Board the free publication *Taking the SAT I–Reasoning Test* and *Registration Bulletin*. These give up-to-date information on fees, locations, test dates, and score reporting. This publication can be obtained by writing to the following address:

> The College Board SAT Program
> P.O. Box 6200
> Princeton, NJ 08541-6200

You can also find out from the College Board how to obtain previously administered *SAT I* tests, if they are not available from your high school.

THE MATHEMATICS SECTION OF THE *SAT I*

The *SAT I* tests only as far as the information found in the basic algebra and geometry courses. At the same time, it tests other aspects of mathematical ability, namely, reasoning and problem solving. Doing well in these areas requires excellent methods of data collection, pattern detection, deductive reasoning, and many other intellectual strategies. Such strategies are not instantly acquired as students learn specific mathematical processes (such as solving a linear equation or finding the length of the third side of a right triangle); rather, they are developed over a long period of time.

Students who receive methodical instruction in reasoning and problem solving exhibit the strongest mathematical skills. In addition, those who take a full sequence of mathematics courses tend to have sharper skills than do students who stop after basic mathematics. As might be expected, the better-trained group tends to score higher on the *SAT I*.

It is true that mathematical ability is a quality that is largely developed as part of your general academic ability; but regardless of how much "inborn talent" students believe they have or the level of mathematics study they have reached, the *SAT I* scores can still fail to be a true measure of a student's ability. Too often, students achieve scores lower than they are capable of achieving because they do not prepare properly. The teacher's task is to bring students as close as possible to their personal ability ceiling (a quantity unknown to all). By following the plan in this book, a teacher can help to ensure that the students' *SAT I* scores will be as high as possible.

The College Board says that students need to know the following topics in these four areas:

- **Arithmetic:** simple addition, subtraction, multiplication, and division; percent; data interpretation (including mean, median, and mode); odd and even numbers; prime numbers; divisibility
- **Algebra:** negative numbers; substitution; simplifying algebraic expressions; simple factoring; linear equations; inequalities; simple quadratic equations; positive integer exponents; roots of numbers; sequences
- **Geometry:** area and perimeter of a polygon; area and circumference of a circle; volume of a box, cube, and cylinder; Pythagorean theorem and special properties of isosceles, equilateral, and right triangles; 30–60–90 and 45–45–90 triangles; properties of parallel and perpendicular lines; simple coordinate geometry; slope; similarity; geometric visualization
- **Other:** logical reasoning; newly defined symbols and operations that are based on commonly used symbols and operations; probability and counting

In addition to a review of these recommended topics, we will present in this book a much more comprehensive review of the material necessary for success on the *SAT I*.

Calculator Use on the *SAT I–Reasoning Test*

A student might think that the advantages of using a calculator on the *SAT I* would be great. In fact, the main advantage afforded is that one can avoid arithmetic errors which would otherwise lower the score in spite of good problem-solving skills.

There are also negative aspects to optional calculator use on the *SAT I*. Test writers have had to reduce the number of questions for which solutions could be found simply by punching data into a calculator. Naturally, these kinds of questions have been replaced with others that may not be as straightforward but are a truer test of mathematical reasoning. For the test taker, other problems arise if he or she

- Is unsure of how to use calculator keys for signed numbers, fractions, percents, and exponents
- Is not adept at handling fractions that convert to repeating decimals
- Cannot quickly change any decimal number—repeating or terminating—back to a fraction to match the multiple-choice answers
- Is simply careless when hitting buttons

Recognizing these pitfalls, this book will illustrate with specific examples how using a calculator can affect the form of a student's answer and will address whether a problem can be solved better with or without a calculator. In addition, the College Board publication *Taking the SAT I–Reasoning Test 1994–1995* offers students some excellent points and cautions regarding calculator use:

- All questions on the test can be answered without a calculator.
- No test questions will require messy or tedious calculations.

- Do not use a calculator on every question. Demonstrate good judgment in determining when and when not to use the calculator.
- Calculators with sophisticated functions are unlikely to provide an advantage.
- Using your calculator too much can cost you time. [The calculator] is meant to aid you in problem solving, not to get in the way.
- Think through how you will solve each problem, then decide whether to use the calculator.

The College Board also reports that students who use calculators "do slightly better than [those] who do not." In this book we will show you how your students can maximize the benefits derived from calculator use and yield results that are more impressive than just "slightly better."

STUDENT-PRODUCED RESPONSE QUESTIONS

Another recent change in the format of the *SAT I* is the inclusion of questions for which students write their own answers. With these, they have a chance to work a problem without being constrained or influenced by multiple-choice answers. Instead, they write answers on a grid by blackening spaces. Some students will probably welcome this feature of the *SAT I*, as they are no doubt accustomed to taking "open-ended" mathematics tests, exams that do not offer a choice of responses. Although there are only ten student-produced response questions on the *SAT I*, the psychological advantage they can provide is great indeed: Because students are unrestricted by a choice of solutions that may not immediately resemble their own, the panic they otherwise may have experienced is reduced on this section. This may allow students to relax a bit and answer this set of questions with a clearer mind.

II. Advising Students on Test Taking

Preparation

Students do not get particularly motivated by teachers admonishing them about the dedicated preparation required for the SAT. This sort of common directive is best left to other forums. Once students realize the importance of a good performance on this test, motivation increases as the test date draws closer. The trick is to harness this motivation earlier, and this can be done by making the instructional program special, whether it is in a separate course or integrated with the regular program. Special instruction generally implies special topics presented in a special mode. This book provides teachers with a host of special topics and suggestions for special modes of presentation. These should be implemented in a manner consistent with a teacher's style and personality.

Test-Taking Procedures

As the test date draws nearer, teachers should begin to discuss with students test-taking procedures that will optimize their performance. There are lots of ways in which a student can take the test. The scheme selected should be carefully thought out, taking into account an individual student's strengths and weaknesses—memory, meticulousness, carefulness, dyslexia, facility with numeral or geometric relations, and so forth—and helping students assess these strengths and weaknesses for themselves. The teacher can only present options. The students must then develop a work plan to suit their self-assessed strengths and weaknesses. One possible work plan would have the student move through items of the test at a steady pace, putting little marks next to each item in the booklet (not on the answer sheet) after it's been dealt with. These marks might be as follows:

- ✓ = definitely correct
- ? = answered but a bit uncertain
- O = omitted, just can't understand on first reading but a second reading might clarify
- ✗ = no idea, come back to it if there is time left

Such a notation system in the margin will allow students to quickly review certain items on their first go-through of the test. Students may choose to review first those items marked "?," then perhaps look at the items marked "O," and so on. The teacher may draw on personal experience to provide other similar test-taking systems. In any case, and regardless of the system selected, some preconceived plan should be in place prior to the actual test. If nothing else, this exercise of planning a strategy will force students to think through the test-taking process, thereby providing the basis for an appropriate state of mind for this crucial time.

To Guess or Not to Guess

Invariably, students seek guidance on the issue of guessing or not guessing on the test. Here, the teacher should take sufficient time to present in a thorough way the statistical thinking that goes into the issue of "correcting for guessing." Before embarking on a discussion of statistics, the teacher must be sure that the class understands the significance (or power) of the concept of probability; that is, they should understand what it means that something is very likely or less likely.

Sometimes a slight digression to a superficial discussion of probability helps convince students that correcting for guessing can really do what it says. The teacher may select dramatic illustrations to exhibit the power of probability—for example, the famous "birthday problem" (see Posamentier and Stepelman, *Teaching Secondary School Mathematics: Techniques and Enrichment Units,* 4th ed., Merrill/Prentice Hall, 1995).

Once students have confidence that probability computations can be believed, the statistical explanation of correcting for guessing would be appropriate, as its validity is based on the strength of probability.

To begin, a teacher may ask the class what grade he should assign on a true–false test to a student who answers incorrectly all 100 questions on the test. Most students will answer, "Zero, of course." It may come as a shock to them when the teacher says "I would probably give him a hundred percent." At this point, students are either perplexed or outraged. The ensuing discussion should focus on the likelihood or probability that getting all questions wrong is equal to that of getting them all correct. Thus, a student would most likely have had to intentionally gotten them all wrong (perhaps as a hoax).

Now to the more realistic example. Once again, a hypothetical test of 100 multiple-choice questions (each item having five choices) should be considered. Suppose the student guesses wildly on all 100 questions (or simply marks responses on answer sheets without looking at the questions). How many of these responses would be expected to be correct? It is anticipated that the class would realize that since the probability of getting any one question right is $\frac{1}{5}$, that on average, $\frac{1}{5}$ of the responses (or 20) ought to be correct responses. On the *SAT I*, the number of points accumulated by correct responses, with a deduction for incorrect answers), determines the test score. The number of points, or raw score, is determined by the formula

$$\text{Raw Score} = R - \frac{W}{n - 1},$$

where R = number of correct responses, W = number of wrong answers, and n = number of choices in each item.

For the sake of illustration, let us consider a student who, having gone through a multiple-choice test (where each item has 5 choices), feels secure that 70 of these items are absolutely correct. She now has a small portion of time remaining in which to process the 30 items which she has not yet answered. If she were to guess wildly on these 30 items (essentially not even looking at questions but just submitting responses), according to probability theory, it would be expected that she would get $\frac{1}{5}$ of the items (or, in this case, 6 items) correct. This would give her 76 answered correctly. Applying the formula above, her raw score would be computed as 76 – 24/4 = 70. Thus, wild guessing neither enhanced nor hurt her score, according to probability concepts; put another way, the chances of helping or hurting her score were the same. This should convince students that wild guessing not only results in no overall gain, but could also cause damage—a rather risky situation to consider!

Let us now return to a student faced with a strategy decision on the remaining 30 items with limited amount of time on the test. Suppose she concentrates her efforts on only 10 of the remaining 30 items and through this extra effort has them correct. This time she omits the remaining unanswered questions (20). Now her raw score would be computed as 80 – 0/4 = 80, since no deduction is made for omissions that are not counted as wrong responses. This strategy resulted in a significantly improved score, and yet may have left the average student a bit concerned with a sizable number of omissions.

Using similar scenarios, a teacher can further convince students that it is wise to develop a strategy for guessing before the test. For example, returning again to the student with her 30 unanswered questions and a limited amount of time on the test, the student, realizing the relatively short amount of time remaining, selects 20 of the remaining questions and by various procedures eliminates 3 choices from each item before guessing the correct answer from the 2 remaining choices. According to probability theory, one-half of these 20 should turn out to be "correctly" answered. The student then omits the remaining 10. Using the above raw score form for this situation, we get 80 – 10/4 = 77.5. This selective guessing results in a worthwhile increase over 70 at the outset of the guessing. Students should be encouraged, based on these and other illustrations, to develop their own test-taking strategies that will help them resolve the questions of "to guess or not to guess," when to guess, and how to guess.

III. NONROUTINE MATHEMATICS TOPICS TO STRENGTHEN STUDENTS' PROBLEM-SOLVING ABILITIES

For decades the acronym *SAT* stood for Scholastic Aptitude Test. The acronym remains the same, but it now represents Scholastic *Assessment* Test. This move away from claiming to measure aptitude seems largely the result of numerous research studies showing that a student can be trained to improve his or her performance on this test. Since aptitude is a trait that supposedly cannot be trained, the test's name has been changed to reflect that conclusion.

In the early days of the *SAT*, the College Entrance Exam Board (CEEB) made strong statements that coaching or training students for this test was fruitless because aptitude was something innate; that is, a student either had high aptitude or did not, and the *SAT* was the instrument that could measure a student's ability. With the change in title, the CEEB now has license to require more mathematics subject matter knowledge than it did before.

A quick overview of early exams showed less reliance on content-dependent problem solving; while this minimized dependence on a knowledge of high school mathematics topics, it nonetheless posed meaningful challenges using topics taken largely from the realm of "general" mathematics. The current exams clearly state which topics/concepts students should know. To minimize reliance on memorization, key formulas and relationships are provided to students on the test. Yes, students can do well on the *SAT I* with only this given material, just as a carpenter can build a house with merely the basic hand tools (saw, hammer, screwdriver, etc.). However, as the carpenter needs to know how to use these basic tools, so, too, must a student know how to use these basic formulas.

When the carpenter is given the advantage of using power tools, his products are completed much more quickly and efficiently. Similarly, a student prepared for the *SAT I* will find that the more powerful her tools, the better her performance ought to be. It is incumbent upon the teacher to provide students with these "power tools."

Let us now look at some ways of providing students with tools that, although presented within high school algebra and geometry courses, are not properly highlighted (or sometimes not even used) as the powerful problem-solving tools they can be.

Using Proportional Relationships to Solve Problems

When the concept of proportion emerges in the school curriculum, it is typically embedded as part of another concept and not seen as an entity deserving its own attention. For example, it is simply *used* during the study of similarity in geometry courses or when considering direct and indirect variation in the algebra course. Yet a problem such as the following is often neglected.

Illustrative Problem

If a apples cost d dollars, what is the price in cents of b apples at the same rate?

Analysis

This problem has been offered to countless groups of students over the past few decades and has caused significant frustration, largely because students frequently do not know how to "grab hold" of the problem. Students who are taught to look for opportunities to apply the concept of proportionality have a relatively easy time with this problem, for they simply set up the following proportion:

$$\frac{a}{b} = \frac{d \text{ dollars}}{c \text{ cents}}$$

Here, the teacher must stress the importance of *keeping the same units of measure in both the numerator and denominator of the fraction* (this is important only for indirect variation, but it is a good habit to get into nonetheless). Consequently, the first fraction here has number of apples in both parts, and the second fraction still requires some modification to get the units the same before any computation can be done. Students can see that the denominator of the second fraction has the unit of measure called for in the statement of the problem and so it is the numerator which must be converted to cents. Students should inductively realize that the numerator can be rewritten as $100d$ cents. The solution to the problem is now obtained by solving the equation:

$$\frac{a}{b} = \frac{100d}{x}$$

$$x = \frac{100bd}{a}$$

To convince them of the "power of proportionality," have students try this problem without giving them any hint; you will probably find a sense of frustration setting in among them as they try to solve it. Many problems on the *SAT I* can be significantly simplified using proportionality techniques.

Simplifying the Procedure to Solve an Equation

Experienced teachers discuss clever techniques to use as shortcuts in simplifying solutions of equations. Some textbooks present these shortcuts as well. For students, any time saved during the test is a welcome relief from the other time pressures of the test. An example of a shortcut technique in solving an equation for x is as follows:

$$\frac{15x^3yz^2}{32y^4} = \frac{25xz^3}{12y^3z}$$

Typically, a student who is called on to solve this equation for x will first cross-multiply and then worry about simplifying where possible. In this case, students can be shown that "canceling through" (i.e., dividing by common factors) the numerator and denominator of each fraction, if possible, is a desirable start. Canceling across numerators and/or denominators is equally legitimate, and will yield a much simpler expression to work with. Please note that our illustration is used to dramatically highlight a point.

Comparing Areas of Similar Figures

Within the scope of the geometry curriculum, students are taught that the ratio of the areas of two (or more) similar figures is equal to the square of their ratio of similitude. Yet beyond simple examples (i.e., trivial squaring exercises), students have little experience with this concept. In preparing students for the *SAT I*, the teacher would be wise to illustrate this technique in a somewhat modified form. Consider the following example.

Illustrative Problem

In the adjoining diagram, the ratio of the diameters of the three semicircles is 2:3:5. Find the ratio of the area of the shaded region to the area of the largest semicircle.

(A) 3:2 (B) 12:25 (C) 13:25 (D) 5:2 (E) 12:13

Analysis

Here, the student should immediately recognize that the ratio of the areas of the three semicircles is $2^2:3^2:5^2$. Since the actual areas are not given, the student has the right (for the purposes of this problem) to let the area of the smaller circle be any convenient number she wishes. Let us assume it to be 4, thereby making the area of the next biggest semicircle 9, and the largest semicircle area 25. In this specific case, the shaded region's area equals 25 - (4 + 9) = 12. Thus, the ratio of the area of the shaded region to that of the largest semicircle is 12:25.

Using area ratios of similar figures for other nontypical problems is a very helpful way of providing students with some extra tools for the *SAT I*. Remember that the ratio of similitude between any two similar figures not only is the ratio of two corresponding sides but also is, in fact, the ratio of any two corresponding linear parts.

Work Problems

A perennial nemesis for algebra students is the "work problem." This problem, typically found in the algebra course, gives the time required for two or more people to do a job working alone and then asks for the time required for the job to be completed with all working together. There are several ways to approach this problem. We shall consider one way here.

Illustrative Problem

Working alone, David can mow a lawn in 2 hours; his father can mow the same lawn alone in 3 hours. How long would it take David and his father working together to mow the entire lawn?

Analysis

In 1 hour, David will have completed 1/2 of the lawn. In 1 hour, working alone, his father will have completed 1/3 of the lawn. If it takes x hours for the two working together, then they will have completed $1/x$ of the lawn in 1 hour. Consequently, the equation $1/2 + 1/3 = 1/x$ will yield the problem's solution.

Although this type of work problem is not found too frequently on the *SAT I*, it can emerge anytime, and the technique for analyzing this kind of problem has a very useful spin-off.

A second kind of work problem, found far less frequently in the algebra course yet found more frequently on the *SAT I*, is the kind of situation where all the workers work at the same rate. This kind of problem opens the door to a useful problem-solving technique that can carry over to other, analogous problems.

ILLUSTRATIVE PROBLEM

A school gymnasium can be painted by 9 painters in 4 days. If 3 painters are unable to work, how long will it take to paint the gym with the remaining workers?

ANALYSIS

Here we introduce a new concept, namely, a measure for the size of the job. This measure, which we will call "painter days" is simply the product of *labor* and *time* to do the job. Here, with 9 painters working 4 days, the size of the job is $9 \cdot 4 = 36$ painter days. With 3 painters missing and x being the time required for the 6 remaining painters to do the entire job, $6x = 36; x = 6$, the time required in the second situation.

Measuring the size of a job can also be useful in the following problem.

ILLUSTRATIVE PROBLEM

If 13 men can cut 364 logs of wood in 14 days working 12 hours a day, how many hours a day must 15 men work to cut 810 logs of wood in 36 days?

ANALYSIS

The size of the second job (cutting 810 logs) is $15 \cdot 36 \cdot h$ (where h is the number of hours per day to do the second job). We can then set up a simple proportion:

$$\frac{13 \cdot 14 \cdot 12}{15 \cdot 36h} = \frac{364}{810}.$$

The solution to this equation, $h = 9$, can be gotten either by using a calculator or by simplifying fractions, remembering that you can cancel by common factors not only the numerator and denominator of the fraction but also across the numerators and denominators.

The teacher should make every effort not only to stress the two important problem-solving schemes illustrated by these work problems but also to highlight the difference between the two.

AVERAGING RATES

It is always curious to notice student reaction to the following question.

Illustrative Problem

If you travel to school at 30 mph and return home along the same route at 60 mph, what is your average speed for the entire trip?

Analysis

If the question is presented by the teacher in the matter-of-fact fashion that would (intentionally) elicit the wrong answer, students will be quick to volunteer the answer 45 mph. They will have gotten this in the expected way of taking a simple average (arithmetic mean) (30 + 60)/2. This is *wrong!*

Once the frustration (or surprise) subsides in the classroom, the teacher might first try to have the students realize that it was "unfair" to weigh both speeds equally when they were "done" for *unequal* amounts of time. A clever student will probably volunteer giving double the weight to 30 mph since it was done twice as long, thus yielding (30 + 30 + 60)/3 = 40, the correct answer.

It may be worthwhile to make a small digression before continuing this development. Our objective here is to convince students of the intuitiveness of this situation: the notion of weighted values.

This average weighting is not always easily understood by students. The teacher may want to provide an illustration that hits closer to home, such as asking students what grade one should get in a course where on 9 of the course tests, a grade of 100 was achieved, and on 1 test a 50 was obtained. Would it be just to say (100 + 50)/2 should determine the students' grades (since the first 9 test scores had an average of 100 and so could be considered as one score of 100)? Here, students will be quick to indicate the unfairness involved, since the grade of 100 was made so much more frequently. The correct way to reach the average would be 950/10 = 95.

What does one do when the situation involves numbers that are not so easily manageable? The commonly taught procedure for handling the situation would be to require the student to find the time going and the time returning, and then take the total distance divided by the total time, which gives the total (or average) rate for the entire round trip. Although this process is perfectly acceptable, it is always in the students' best interest to show them the most efficient procedure. Here we can use a concept from "higher mathematics" known as the *harmonic mean*: the reciprocal of the arithmetic mean of the reciprocals of the rates to be averaged. In the above illustration, we can apply this harmonic mean (for rates a and b), written in simplified form as $(2ab)/(a + b)$, which gives us

$$\frac{2 \cdot 30 \cdot 60}{30 + 60} = 40.$$

The harmonic mean can be used to average 3 or more rates, provided the base (in this case the distance) is the same for all the rates.

The simplified form for the harmonic mean of three rates a, b, and c would be

$$\frac{3abc}{ab + ac + bc}.$$

For the four rates a, b, c, and d, the harmonic mean can be found with the simplified form

$$\frac{4abcd}{abc + abd + acd + bcd}.$$

The rates to be averaged need not be rates of speed; they could be rates of other things, such as rate of purchase (i.e., price per item). Consider the following.

ILLUSTRATIVE PROBLEM

Jill buys $30 worth of each of two different kinds of pens, priced at 40 cents and 60 cents per pen, respectively. What is the average price in cents paid for each pen?

ANALYSIS

The base over which both rates of purchase are made, $30, is the same. Hence, we apply the harmonic mean formula $(2ab)/(a + b)$ and we get $(2 \cdot 40 \cdot 60)/(40 + 60) = 48$ cents per pen.

This procedure of finding the average rates is rarely shown to high school students, yet it is very simple to apply and can save significant amounts of time on the *SAT I*. It is highly desirable, therefore, to introduce it to students and give them ample time to practice using it.

FINDING AREAS OF "NAMELESS" GEOMETRIC FIGURES

Most geometric figures that have names (such as square, triangle, pentagon) also have formulas for finding their area. Most of these are presented in the course of high school mathematics instruction. The *SAT I* occasionally requires finding the area of odd-shaped figures or finding the areas of parts of figures whose shapes can be described only as "the shaded region." Problems of this sort are difficult to categorize and even more difficult to generate solutions for. The teacher can merely suggest a number of approaches for attacking this sort of problem with the hope that one of these will lead to a successful solution. One possibility for solving this sort of problem is to *divide the sought-after region into smaller identifiable parts,* and find the areas of each of the smaller parts, where the sum or difference would lead to the desired solution. Consider, for example, the following problem.

ILLUSTRATIVE PROBLEM

In the adjacent diagram, find the area of the shaded region formed by semicircles with diameters of 8 and 4, respectively, in the circle of diameter 12.

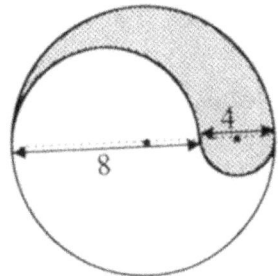

ANALYSIS

This tear-shaped region has no official name, and there is no formula readily available for finding its area. To find its area, we need to find the area of the large semicircle (diameter 12), subtract from it the area of the smaller semicircle (diameter 8), and then add to it the area of the small semicircle (diameter 4):

$$\frac{1}{2}\pi 6^2 - \frac{1}{2}\pi 4^2 + \frac{1}{2}\pi 2^2 = 12\pi.$$

Teachers would be wise to allow students to practice on a variety of these nameless-region areas, since practice is probably the best way to prepare for this kind of question.

LIKELIHOOD PROBLEMS

There are a number of problems on the *SAT I* that rely not on formal topic training in mathematics but rather on a thinking or analytical skill the student might possess. One such problem requires students to consider extreme possibilities.

ILLUSTRATIVE PROBLEM

In a dark room where colors are not distinguishable, how many socks must a person take from a drawer containing 12 blue socks, 8 black socks, and 6 green socks, to be assured of having two socks of the same color?

ANALYSIS

For this kind of problem, the student must assume the "unluckiest" possibility, since only then will she be *assured* of success. In the unlucky situation, the person would pull 3 socks each of a different color. On his fourth draw, he would have to pull a sock that matches one of the previous three colors. Thus, the problem's answer is 4 socks, and yet students will say, "But he could have pulled a matching pair of socks on the first two tries." The teacher must stress that, although this is possible, it is only when 4 socks are drawn that a match is *guaranteed*.

A similar kind of problem requiring slightly different thinking follows:

ILLUSTRATIVE PROBLEM

In a dark room where colors are not distinguishable, how many socks must a person take from a drawer containing 12 blue socks, 8 black socks, and 6 green socks, to be assured of getting one sock of each color?

ANALYSIS

Once again, we must assume an unlucky possibility, which would have him drawing 12 socks of the same color (blue) and then 8 socks of the same color (black), thus not guaranteeing one sock of each color even in the first 20 draws. However, since the first two colors are exhausted, on draw 21 he must pick a sock of the third color.

Once again, it should be made clear what it means to be *assured* of an event taking place, since it is possible to pull one sock of each color with the first three tries. This sort of reasoning is an important part of *SAT I* preparation.

SUCCESSIVE DISCOUNTS

For reasons unclear to us, percentage problems have long been a nemesis to most students. Problems get particularly unpleasant when multiple percents need to be processed in the same problem. Consider the following problem:

ILLUSTRATIVE PROBLEM

Wanting to buy a coat, Howard is faced with a dilemma. Two competing stores next to each other carry the same brand coat with the same list price, but at two different discount schemes. Store A offers a 10% discount year round on all its goods, but on this particular day offers an additional 20% on top of their already discounted price. Store B simply offers a discount of 30% on that day in order to stay competitive. How many percentage points difference is there between the two options open to Howard?

ANALYSIS

At first glance, students will assume there is no difference in price, since 10 + 20 = 30, yielding the same discount in both cases. The clever student will see that this is not correct, since in store A only 10%

is calculated on the original list price, with the 20% calculated on the lower price; while at store B the entire 30% is calculated on the original price. Now, the question to be answered is, what difference in percent is there between the discount in store A and store B?

The expected procedure will probably have the student assume the cost of the coat to be $100, calculate the 10% discount, yielding a $90 price; an additional 20% of the $90 price will bring the price down to $72. In store B, the 30% discount on $100 would bring the price down to $70, giving a discount difference of $2, or in this case, 2%. This procedure, although correct and not too difficult, is a bit cumbersome and does not always allow full insight into the situation.

A more interesting procedure for finding the single percent of discount/increase equivalent to two or more successive discounts/increases would be as follows:

(1) Change each of the percents involved into decimal form:
.20 and .10
(2) Subtract each of these decimals from 1.00:
.80 and .90 (for an increase, add to 1.00)
(3) Multiply these differences:
$.80 \cdot .90 = .72$.
(4) Subtract this number (i.e., .72) from 1.00:
$1.00 - .72 = .28$ (if the result of step 3 is greater than 1.00, subtract 1.00 from it to obtain the percent of increase).

When we convert .28 back to percent form, we obtain 28%, the equivalent of successive discounts of 20% and 10%. This percent (i.e., 28%) differs from 30% by 2%.

In addition, successive increases, combined or not combined with a discount, can also be accommodated in this procedure by adding the decimal equivalent of the increase to 1.00 where the discount was subtracted from 1.00 and then continuing the procedure in the same way. (For an in-depth discussion of this procedure, see Posamentier and Stepelman's *Teaching Secondary School Mathematics: Techniques and Enrichment Units* [4th ed., Merrill/Prentice Hall, 1995], p. 304.)

This procedure not only streamlines a typically cumbersome situation but also provides some insight into the overall picture. For example, let's look at this question: Is it advantageous to the buyer in the above problem to receive a 20% discount and then a 10% discount or the reverse, 10% discount and then a 20% discount? The answer to this question is not immediately intuitively obvious. Yet, since the procedure just presented shows that the calculation is merely multiplication, a commutative operation, we find immediately that there is no difference between the two. As always, practicing with different arrangements and variations on this theme will help verify its use on the *SAT I*.

Clock Problems

The face of a clock presents a number of interesting problems sometimes found on the *SAT I*. Its modular nature lends the clock very nicely to a rather "automatic" technique for dealing with many problems involving the face of a clock.

ILLUSTRATIVE PROBLEM

At what time after 4:00 will the minute hand overtake the hour hand?

(A) 4:20 (B) 4:21 (C) $4:21\frac{7}{11}$ (D) $4:21\frac{9}{11}$ (E) 4:22

ANALYSIS

We may deal with this problem as we would if asked to find the time it took a fast car to overtake a slower one. Let us speak of the rate (i.e., speed) of the hour hand as r. Then the rate of the minute hand is $12r$. The distance the hands travel will be measured by the minute marks on the clock. What we seek in this problem is the distance (in minutes) the minute hand must travel to overtake the hour hand.

Let this distance be x. Therefore, the distance that the hour hand must travel is $x - 20$, since at 4:00 it has a 20-minute head start over the minute hand. Since time equals distance divided by rate, the time the minute hand travels is $x/(12r)$, and the time that the hour hand travels is $(x - 20)/r$. Since both hands travel the same amount of time,

$$\frac{x}{12r} = \frac{x - 20}{r}$$

$$x = 12(x - 20) = 12x - (12)(20)$$

$$11x = (12)(20)$$

$$x = \frac{12}{11}(20)$$

$$x = 21\frac{9}{11}$$

Therefore, at $4:21\frac{9}{11}$, the minute hand will overtake the hour hand. The answer is choice D.

Note: In the relation $x = (12/11) \cdot 20$, the quantity 20 is the number of minutes of head start that the hour hand had over the minute hand. The ratio 12/11, then, is the number of minutes required, per minute of head start, for the minute hand to overtake the hour hand. Therefore, we can substitute any known value for the 20, and find the time required by multiplying by 12/11. For example, if the time at the start is 8:00, the hand overlap is found by finding the number of minutes, x, the minute hand must travel after 8:00 to reach the hour hand. Thus

$$x = \frac{12}{11} \cdot 40 = 43\frac{7}{11}.$$

The minute hand will overtake the hour hand at $8:43\frac{7}{11}$.

Another way of looking at this relation is to say that the head start of number of minute markers is the time it would take the minute hand to reach the hour hand if the hour hand did not move. But because the hour hand does move, it takes 12/11 as long.

A "mechanical" technique for solving problems of this type is to realize that various hand positions repeat every $1:05\frac{5}{11}$. Therefore, starting at noon, when we know that the hands of a clock overlap, we can simply add the appropriate multiple of $1:05\frac{5}{11}$ to get the correct overlap times. For example, to find out what the next overlap will be after 4:00, we add 4 times $1:05\frac{5}{11} = 4:20\frac{20}{11}$ to get $4:21\frac{9}{11}$. This idea can be applied to many variations of the clock problem.

SEQUENCE PROBLEMS

Invariably, when one thinks of reasoning problems, the first kind of problem that comes to mind is that of finding the pattern in a sequence of numbers and then either finding a missing number or the succeeding numbers of the sequence. Unfortunately, there is no one technique for attacking these problems. We will simply present a series of possible problems with the hope that the teacher will be able to create additional samples to prepare students for them when they appear on the *SAT I*.

ILLUSTRATIVE PROBLEM

Find the next number in the following sequence:

2, 5, 9, 14, 20, ___

ANALYSIS

A first attempt would be to see if there is a common difference or common quotient between the numbers. The successive differences between these numbers are 3, 4, 5, 6. The pattern clearly dictates that the next difference should be 7. Thus, the next number in the series is 27.

ILLUSTRATIVE PROBLEM

Find the next three terms in the following sequence:

0, 1, 2, 3, 6, 7, 14, 15, 30, __, __, __.

Analysis

A quick inspection shows no common difference between terms. However, embedded in the sequence are several pairs of consecutive numbers—(0, 1), (2, 3), (6, 7), (14, 15)—which should begin to give a clue about the pattern working backwards. One notices that each pair begins with the number that is twice its predecessor: 2 is twice 1, 6 is twice 3, 14 is twice 7, and 30 is twice 15. Thus, the next number after 30 would be its successor, 31, which we double to get 62 and then add 1 to get 63. Thus, the next three terms of the given sequence are 31, 62, 63.

Illustrative Problem

Find the next number in the sequence 2, 9, 23, 44, 72, ___.

Analysis

In the search for common differences, we find the differences to be 7, 14, 21, 28. The pattern obviously then extends to 35, which when added to 72 gives us the next number of the sequence, 107.

Sometimes the sequence of numbers is further confused by a pattern that does not steadily increase or decrease. Consider the following problem:

Illustrative Problem

One term does not belong to the following sequence:

1, 8, 3, 6, 9, 4, 7, 2.

Which of the following numbers should replace the incorrect term?

(A) 9   (B) 6   (C) 2   (D) 5   (E) 6

Analysis

The up-and-down fluctuation of this sequence *could* signal two sequences interspersed. Hence, we will look at every other number to see if there is a pattern. The even-position terms build a sequence: 8, 6, 4, 2. This seems to be a complete and correct sequence. An inspection of the odd-positioned terms shows the sequence: 1, 3, 9, 7. If the 9 were replaced by 5, we would have consecutive odd numbers. Thus, choice D is the correct response.

A teacher preparing students for this sort of test item would be wise to invent as many variations as possible so as to help students anticipate this item on the test.

COUNTING PROBLEMS

The topic of counting, or, to be more specific, systematically counting, is finding a more prominent role in the curriculum. It is usually combined with the study of permutations, combinations, and probability, but it is used on the *SAT I* to determine if a person can organize a situation to demonstrate some sophisticated counting methods.

ILLUSTRATIVE PROBLEM

In a room with 6 people, each person shakes hands each of the 5 other people exactly once. How many handshakes are there?

ANALYSIS

Students may draw a diagram representing six people, and draw lines to indicate handshakes. This could lead to a successful counting procedure if done systematically.

However, this particular problem can be solved by counting in the following way. The first person shakes hands with 5 others. The second person, having shaken the first person's hand, shakes hands with 4 other people. The third person shakes hands with 3 other people, the fourth person with 2 others, and the fifth person with 1 other person, making a total of 15 handshakes. (Note: The sixth person is not considered, since she already has shaken everybody else's hand.)

Systematic counting can also be seen in the following problem.

ILLUSTRATIVE PROBLEM

A hotel issues parking permits consisting of 2 letters (not including O) followed by 3 numerals with no numeral repeated. How many different permits can it print?

ANALYSIS

We will look at the number of permits by systematically counting the number of possible permit numbers. Consider the 5 positions to be filled. The first 2 can each be filled in any one of 25 different ways, since there are 25 letters to choose from. (Remember, O is excluded.)

The next position can be filled in any one of 10 different ways, since there are 10 digits to choose from. With no digit permitted to be repeated, the fourth place can be filled in any one of 9 ways, and the fifth position can be filled in any one of 8 ways. To find the number of permits possible, we multiply $25 \cdot 25 \cdot 10 \cdot 9 \cdot 8$, which gives us 450,000.

Counting can sometimes require nothing more than a little organization. Consider the following problem.

ILLUSTRATIVE PROBLEM

In numbering the pages of a book, 3,357 digits are required. What is the number of pages in the book?

(A) 998   (B) 3,357   (C) 1,091   (D) 1,119   (E) 1,116

ANALYSIS

Students should realize that we must count these digits in groups of similar types of page numbers, using the following scheme:

Pages 1–9 use 9 digits.
Pages 10–99 use 90 · 2 digits, or 180 digits.
Pages 100–999 use 900 · 3 digits, or 2,700 digits.

At this point, 2,700 + 180 + 9 = 2,889 digits of the 3,357 digits that have been used to number pages 1–999.

Remaining are 468 digits, which can be used to make 117 four-digit numbers. Therefore, the total number of pages is 999 + 117 = 1,116.

COMPARING FRACTIONS

We leave for last the topic of comparing fractions. Today's *SAT I* no longer includes an item calling for the simple comparison of fractions, as was done on prior examinations. On today's *SAT I* the student can easily dispose of such questions by using a calculator. This makes the task trivial. Either they can simply compare (by observation) the decimal equivalents of each of the fractions being compared or use the more general relationship for comparing fractions:

$$\frac{a}{b} > \frac{c}{d} \text{ if and only if } ad > cb.$$

This becomes particularly useful for the quantitative comparison questions on the *SAT I*.

ILLUSTRATIVE PROBLEM

Which of these two expressions is larger (if either)?

$$\frac{m^2 + n^2}{m + n} \text{ or } m + n, \text{ where } m \text{ and } n \text{ are integers greater than } 1$$

ANALYSIS

One of the techniques for solving this problem is to have the students substitute a small, manageable number in place of the m and n to see how the two terms compare. If we let $m = 2$ and $n = 3$, then the first expression $\frac{m^2 + n^2}{m + n} = \frac{4 + 9}{2 + 3} = \frac{13}{5} = 2\frac{3}{5}$, as compared to $2 + 3 = 5$.

The general case, using the above relationship, is useful for a student to know when the problem becomes more complicated or does not lend itself to this substitution method. There we want to inspect the "cross-products" of the two fractions:

$$\frac{m + n}{1} \; ? \; \frac{m^2 + n^2}{m + n}$$

That is: $(m + n)(m + n) \; ? \; (1)(m^2 + n^2)$.

We know that $m^2 + n^2 + 2mn > m^2 + n^2$, since $2mn > 1$. Therefore, $m + n$ is the greater of the two expressions.

ILLUSTRATIVE PROBLEM

Which of the following fractions is greater (or are they equal, or can it not be determined)?

$$\frac{m}{n} \text{ or } \frac{m + k}{n + k}, \text{ where } m, n, \text{ and } k \text{ are positive}$$

ANALYSIS

Here the substitution method will not be easy to use, for the selection of a proper or an improper fraction for m/n will yield different results. To gain a full understanding of this question, the fraction comparison method shown above should be used. This fraction comparison relationship tells us

$$\frac{m}{n} \; > = < \; \frac{m + k}{n + k}$$

if $mn + mk \; > = < \; mn + nk$, which requires comparing mk and nk. These cannot be compared with the information given; therefore, the original two fractions cannot be compared.

The topics and techniques presented in this section are not so much problem-solving techniques as they are enhancements to a student's understanding of mathematics; the problem-solving strategies in Part IV are more geared to view the heuristics, or strategies, of the problem-solving process. Together, these two sections give teachers (and ultimately students) the bulk of the material and ideas required to prepare the test taker as well as possible for this test. Naturally, the teacher must seek other examples of these topics and techniques to properly "plant" this new material in the students' mathematics domain. We are merely presenting a relatively small number of illustrative examples as a model.

IV. Specific Problem-Solving Strategies

Strategy 1: Working Backwards

On the surface, the title of this problem-solving method sounds confusing. This stems from the unnaturalness of the procedure. From earliest times, students typically are taught to solve problems in a straightforward way. This is the way typical mathematics textbook problems are intended to be solved. Unfortunately, a substantial portion of this supposed "problem solving" is rote. Students may struggle through one problem in a section but then the remainder of the problems are usually very similarly solved, requiring little imaginative thinking on the part of students. When, in the high school geometry course, students are required to do proofs, they once again look for ways in which they can merely repeat previous rote procedures to solve successive problems. These exercises are not true mathematical problem-solving experiences but merely repetitions of previously used methods.

When students are confronted with a test such as the *SAT I*, they are no longer faced with a series of repetitious problems; rather, they are presented with problems that require thought and reasoning. To address these problems, teachers tend to embark on a venture to teach problem solving where, finally, the emphasis is brought to the *process* of problem solving and not just the *product*, the correct answer. Even in this more productive setting, teachers tend to pay too much attention to the repetitiveness of similar kinds of procedures to solve problems, with the notion that practice makes perfect. In this book we are trying to focus more attention on the process. Repetition of a particular type of problem will distract from the process and focus attention on the type of problem being repeated. This is particularly true when we look at the technique of working backwards, a procedure, as stated previously, used subconsciously and yet never really addressed head on. It deserves special attention because of its usefulness in so many varied settings.

When we look at the procedures students are shown in many of their typical textbook exercises, there are some very useful techniques used but unfortunately taken for granted. Students are often required to subconsciously reason in the reverse order. An obvious example of working backwards is the procedure students should use when doing proofs in the high school geometry course. They should begin with what they are looking to prove before doing anything else. A clever problem solver then begins to work backwards from this desired conclusion to a point where the given information is utilized. The proof can then be written directly from there.

In everyday life, one uses this sort of reasoning when, for example, one is planning a trip to go from point A to point B. One does not merely plan the path of the trip exclusively from point A and search for the various routes that lead to point B. If B is a large city it is often best to plan a route straightforwardly; however, if B is a small city with few access roads and A is a large city, it will almost always be vastly superior to plot a route from B to A, as the plotter will inevitably reach an obvious access road. Even on the *SAT I*, when students respond to multiple-choice questions, they are well advised to inspect the choices, the type of choices, and their approximate magnitude before embarking on the problem's solution.

Although most problems require, at least to a very small extent, some reverse reasoning, there are some problems the solutions to which are dramatically facilitated by working backwards. Consider the following problem.

ILLUSTRATIVE PROBLEM

The sum of two numbers is 2. The product of the same two numbers is 3. Find the sum of the reciprocals of these two numbers.

(A) $\dfrac{2}{3}$ (B) 1   (C) $1 \pm i\sqrt{2}$ (D) $2 \pm 2i\sqrt{2}$ (E) $\dfrac{3}{2}$

ANALYSIS

Most students immediately generate the equations $x + y = 2$ and $xy = 3$. They are taught to solve these equations simultaneously by substitution. If in this complicated example they don't make an algebraic mistake, they will come up with rather unpleasant looking values for x and y. They then need to find the reciprocals and then their sum. This rather complicated solution can be made much simpler by starting from the desired result, namely $1/x + 1/y$.

The student then asks, "Where might this have come from?" or "What can I do with these fractions?" One possibility is to find their sum, which is $(x + y)/xy$.

Working backwards further to the given information, we find that this fraction must be 2/3 (choice A), since the numerator, $x + y$, is 2 and the denominator, xy, is 3.

Although this problem is probably more complicated than a typical *SAT I* question, it very nicely dramatizes the benefit of using this reverse problem-solving procedure. It is important to stress that most mathematics problems employ some degree of reverse reasoning. Sometimes the reverse reasoning looks a little different. Consider the following problem.

ILLUSTRATIVE PROBLEM

A tennis tournament has 32 players. One match loss eliminates a player. How many matches must be played (or defaulted) to get a single winner?

(A) 16   (B) 30   (C) 31   (D) 32   (E) None of these

ANALYSIS

Students tend to simulate a possible tournament beginning with the first matches and the successive eliminations and counting their hypothetical case. A much more efficient method has students looking at the end result: the winner. Working backwards, one may ask how many losers must there have been? 31. How do we get to 31 losers? 31 matches! And so the problem is solved.

Students should be alerted to problem situations in which the final or end conditions are given and the original or starting conditions are asked for. These problems usually lend themselves quite readily to the "working backwards" strategy of problem solving. The following problem illustrates this situation quite vividly.

ILLUSTRATIVE PROBLEM

In Round 1 of a game, Pat gave Randi and Theresa as much money as they each had.
In Round 2, Randi gave Pat and Theresa as much money as they each had.
In Round 3, Theresa gave Pat and Randi as much money as they each had.
At this point, they each had $8.00. How much money did Randi start with?

(A) $24   (B) $13   (C) $7   (D) $4   (E) None of these

ANALYSIS

Most students will probably begin by trying to set up a system of three equations in three unknowns. Can it be done? Yes! And you should take them through the exercise. However, unless they are extremely adept and careful when working with parentheses, the final equations are likely to elude them. Even if they do arrive at the correct set, the equations must be solved simultaneously in pairs, a tedious process at best. Let's examine the process a bit and see where the equations take us:

Round	Pat (x)	Randi (y)	Theresa (z)
1	$x - y - z$	$2y$	$2z$
2	$2x - 2y - 2z$	$3y - x - z$	$4z$
3	$4x - 4y - 4z$	$6y - 2x - 2z$	$7z - x - y$

Thus we arrive at the following system of equations:

$$4x - 4y - 4z = 8$$
$$-2x + 6y - 2z = 8$$
$$-x - y + 7z = 8$$

Which leads to $x = 13, y = 7,$ and $z = 4$.

Point out to your students that the ending situation (they each have $8) is given and the starting situation (how much Randi started with) has been asked. Now apply the working-backwards strategy:

	Pat	Randi	Theresa
End of round 3	8	8	8
End of round 2	4	4	16
End of round 1	2	14	8
Start	13	7	4

The correct answer is (C) $7.

The teacher would be wise to highlight and make students aware of the importance and prevalence of this reverse-reasoning procedure in problems heavily dependent on it as well as in problems where this approach may only subtly appear to help find a solution. Students should not only be made aware of alternate problem-solving procedures but also be urged to independently seek alternate solutions to problems they have solved. The teacher ought to encourage students not to be conclusively satisfied with a successful solution to a problem; rather, they should use this "success" as a motivation to search for other methods of solution. Often, the reverse strategy for solving a problem can present just such an alternate path to a solution. Through this sort of reinforcement, this rather unnatural problem-solving procedure will take hold and become an integral part of the student's arsenal of problem-solving procedures.

STRATEGY 2: FINDING A PATTERN

A famous mathematician was once asked to define mathematics. He replied, "Mathematics is a search for patterns!" Such a statement acknowledges the important role that patterns play in all mathematics and especially in problem solving. Indeed, many of the problems on the *SAT I* can be resolved if the students can recognize a pattern within a given set of data. Thus, a great deal of your class time should be spent in practicing pattern recognition. This should be done not only via formal instruction but also informally as well, whenever patterns occur. Begin with simple arithmetic or geometric sequences, such as 3, 7, 11, 15, ___, ___; or triangle, rectangle, pentagon, hexagon, ___, ___. Then move into sequences in which the pattern is not so obvious. Have your students verbalize the pattern rule they have discovered. This helps to fix the pattern sequence in their minds, to help them recognize it the next time it appears.

ILLUSTRATIVE PROBLEM

Which of the following pairs of numbers would occur consecutively between 21 and 377 in the given sequence:

$$3, 5, 8, 13, 21, \ldots, 377$$

(A) 31, 43   (B) 144, 233   (C) 56, 72
(D) 171, 296   (E) None of these

Analysis

The pattern in this sequence is not at all obvious when students encounter it for the first time. Since it often occurs in problems, you should carefully introduce them to such sequences, known as Fibonacci sequences, in which each term beginning with the third term is the sum of the two previous terms. (The sequence 1, 1, 2, 3, 5, . . . following this pattern is often called *the* Fibonacci sequence.) Students should recognize that each term after the first two is found by adding the previous two terms. That is, $3 + 5 = 8$, $5 + 8 = 13$, $8 + 13 = 21$, and so on. If we now continue the sequence, we obtain the following:

3, 5, 8, 13, 21, 34, 55, 89, 144, 233, 377.

The missing terms would be those in choice B, 144 and 233.

Illustrative Problem

If the given figures were extended down, following the same pattern, how many black squares would there be in the first 20 rows:

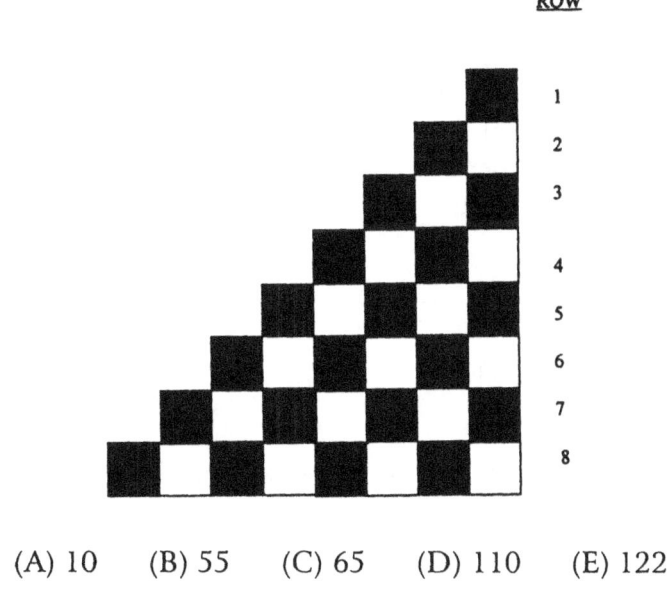

(A) 10   (B) 55   (C) 65   (D) 110   (E) 122

Analysis

Have your students count the black squares in the first eight rows shown. We obtain the sequence 1, 1, 2, 2, 3, 3, 4, 4, ... We can assume that this pattern will continue, so that in the 20th row (as in the 19th row) there will be 10 black squares. Now the problem has been reduced to finding the sum of two interspersed sequences of 1, 2, 3, 4, . . . , 10. We can add these directly, or we can note that $1 + 10 = 11$, $2 + 9 = 11$, $3 + 8 = 11$, and so on. This gives us $5 \cdot 11$, or 55. Since we require the use of two such sequences, the total number of black squares will be 2 · 55, or 110, choice D. Notice that, while a calculator might be used to find the sum, it would probably be more time consuming. However, some of your students might feel a bit more secure doing this.

The pattern sequences in the previous two problems were somewhat obvious. However, patterns often occur when least expected. You must encourage your students to become sensitive to the possible appearance of a pattern whenever they are solving a problem. Recognizing that a pattern exists often makes the solution much simpler.

ILLUSTRATIVE PROBLEM

Which of the following represents the cube of 6,456?

(A) 632,944,312   (B) 74,252,334,444   (C) 592,866,468,204
(D) 269,085,666,816   (E) 788,486,666,122

ANALYSIS

While some students may actually try to find the cube of 6,456 (it can be done on a calculator if the display is large enough), you should emphasize that it can save them time and ensure accuracy if they look for a pattern. That is, the units digit for powers of many numbers form a pattern. These should be practiced, studied, and recognized. Students probably already know that the units digit of the powers of 5 will always be 5. They may also know that the units digits of the powers of 7 form a sequence of four repeating terms:

7	1680_7_	4035360_7_
4_9_	11764_9_	28247524_9_
34_3_	82354_3_	197732674_3_
240_1_	576480_1_	1384128720_1_

In this particular problem, students should recognize that the units digit of the powers of 6 will always be a 6. Thus, any power of 6,456 must also have a units digit of 6. The correct answer can only be (D) 269,085,666,816, since it is the only choice ending in 6.

ILLUSTRATIVE PROBLEM

How many angles are formed by 10 distinct rays with a common endpoint no two of which lie on the same line?

(A) 45   (B) 35   (C) 26   (D) 18   (E) None of these

ANALYSIS

Students can actually draw a large, accurate diagram to answer this question. However, the drawing soon becomes rather confusing as we lose track of the angles we are forming. Let's start with a simpler case, and see if a pattern emerges as we increase the number of cases.

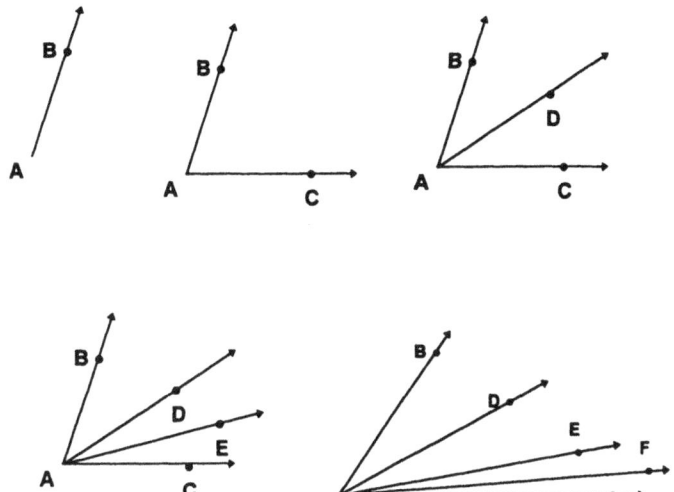

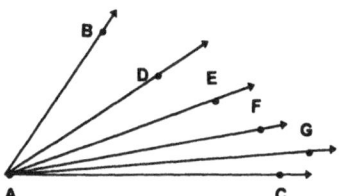

Let's summarize this data in a table:

Number of rays	1	2	3	4	5	6	7	...
Number of angles formed	0	1	3	6	10	15	21	...

The students should not have to draw more than the case of 5 or 6 rays in order to see the emerging pattern:

0, 1, 3, 6, 10, ...

This is a sequence whose *differences* form the simple arithmetic progression 1, 2, 3, 4, 5, ... Continuing this sequence to 10 terms is simple:

0, 1, 3, 6, 10, 15, 21, 28, 36, <u>45</u>.

The correct answer is (A) 45.

Note: These numbers are referred to as the "triangular" numbers because they can be represented geometrically in a triangular array:

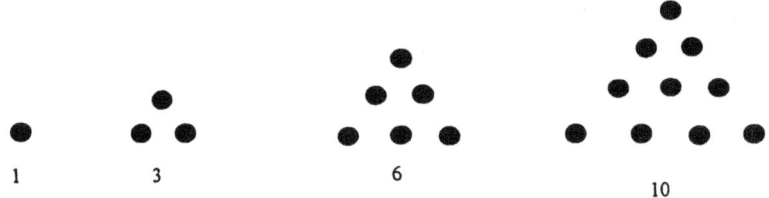

They tend to appear quite often in mathematical problem situations.

ILLUSTRATIVE PROBLEM

Find the sum of the following series:

$$\frac{1}{1 \cdot 2} + \frac{1}{2 \cdot 3} + \frac{1}{3 \cdot 4} + \ldots + \frac{1}{99 \cdot 100}$$

(A) $\dfrac{1}{9900}$ (B) $\dfrac{99}{100}$ (C) $\dfrac{87}{265}$
(D) $\dfrac{42}{990}$ (E) None of these

ANALYSIS

It is true that students can actually compute the individual values for each of the fractions and then add: $1/(1 \cdot 2) = 1/2$, $1/(2 \cdot 3) = 1/6$, $1/(3 \cdot 4) = 1/12, \ldots, 1/(99 \cdot 100) = 1/9900$, the final answer being the sum of $1/2 + 1/6 + 1/12 + 1/20 + \ldots + 1/9900$. Obviously, this would be a rather laborious and difficult task, even with a calculator. Let's see, then, if there is a pattern we can use instead.

$$\frac{1}{1 \cdot 2} = \frac{1}{2}$$

$$\frac{1}{1 \cdot 2} + \frac{1}{2 \cdot 3} = \frac{2}{3}$$

$$\frac{1}{1 \cdot 2} + \frac{1}{2 \cdot 3} + \frac{1}{3 \cdot 4} = \frac{3}{4}$$

$$\frac{1}{1 \cdot 2} + \frac{1}{2 \cdot 3} + \frac{1}{3 \cdot 4} + \frac{1}{4 \cdot 5} = \frac{4}{5}$$

Thus the pattern emerges that strongly suggests that the sum of this series, with its last term of $1/(99 \cdot 100)$, will be $99/100$, choice B.

ILLUSTRATIVE PROBLEM

Find the sum of the first 30 odd numbers.

(A) 450   (B) 240   (C) 600   (D) 900   (E) None of these

ANALYSIS

Although this problem can easily be solved by using the formula for the sum of the terms of an arithmetic progression [$S = \frac{n}{2}(a + l)$, where n is the number of terms, a is the first term, and l is the last term], the student must first determine what the first 30 odd numbers are (i.e., 1, 3, 5, 7, ..., 59). This is an easy place to make a simple arithmetic error.

Some of your students may even decide to use their calculators and plug in $1 + 3 + 5 + 7$ and so forth, until 30 odd numbers have been summed. However, let's examine a simpler version of the problem and see if we can recognize a pattern that could help us solve the original problem.

Begin with the simplest version of all, one single odd number (1), then two odd numbers (1 + 3), and so on, to see if we discern a pattern. Have your students make a table to keep track of their results.

Odd numbers	Number of odd numbers	Sum
1	1	1
1 + 3	2	4
1 + 3 + 5	3	9
1 + 3 + 5 + 7	4	16
.	.	.
.	.	.
.	.	.
1 + 3 + 5 + 7 + ... + n	n	n^2

The problem is now easily solved. The sum of the first 30 odd numbers will be 30^2, or 900, which is choice D.

ILLUSTRATIVE PROBLEM

A circular portion of the school yard has been set aside for the different grades to plant their gardens. The circle is divided up into different-size regions by stretching ropes across the circle. What is the maximum number of regions that can be obtained by using 7 ropes?

(A) 7   (B) 14   (C) 22   (D) 29   (E) None of these

ANALYSIS

We can solve this problem rather easily if the students recognize the situation and can recall the formula for the number of regions as $y = \frac{x^2 + x + 2}{2}$, where y is the maximum number of regions obtained with x ropes.

But since it is rather unlikely that your students will recall this formula, we need another method. Students might try drawing the situation with 7 chords, but this would not guarantee that they would have found the *maximum* number of regions called for in the problem. However, let's reduce the problem to simpler cases, keep track as we expand them, and look for a pattern to emerge.

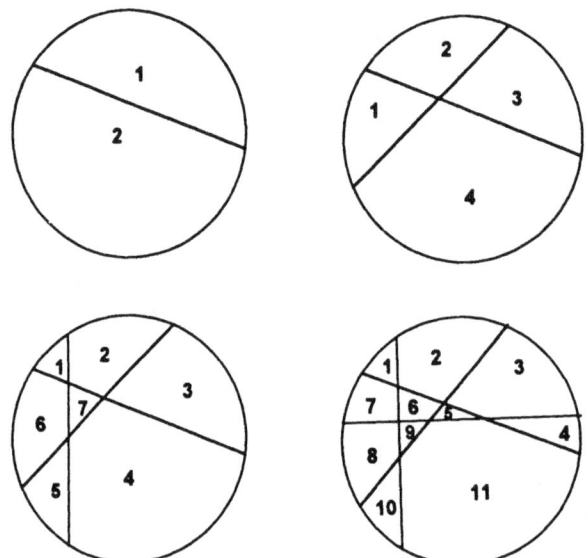

We can make use of a table to observe our results:

Number of ropes	0	1	2	3	4	...
Maximum number of regions	1	2	4	7	11	...
Differences		1	2	3	4	5

We can now continue the process of making drawings and counting the regions—or, even better, we can simply make use of the pattern that appears in the table and extend the table until we have seven ropes. The number of regions will be 29, choice D.

The use of patterns is a most important strategy in problem solving. In fact, pattern recognition is often combined with other strategies in order to resolve problem situations. This will be apparent as we proceed through other strategies in this section. A word of caution is here in order. Not all patterns lead to obvious generalizations. For example, the sequence 1, 2, 4, 8, 16 does *not necessarily* imply that the next term is 32, since 31 also "works." That is, 1, 2, 4, 8, 16, 31, 57, 99, ... is a perfectly legitimate sequence representing the number of regions into which a circle is partitioned by a progressively increasing number of points on a circle (joined by straight lines).

ILLUSTRATIVE PROBLEM

The first four figures in a tile pattern are shown below. How many squares must be added to the fifth figure to make the sixth figure in the series?

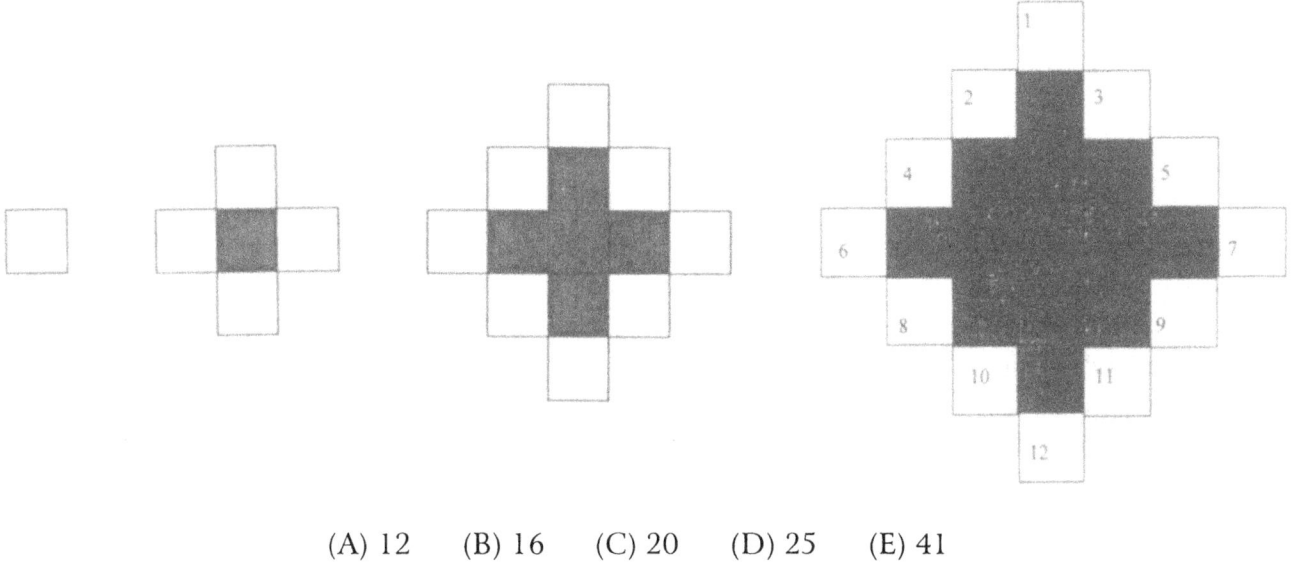

(A) 12   (B) 16   (C) 20   (D) 25   (E) 41

ANALYSIS

Your students can solve the problem directly by drawing the next two figures in the series.

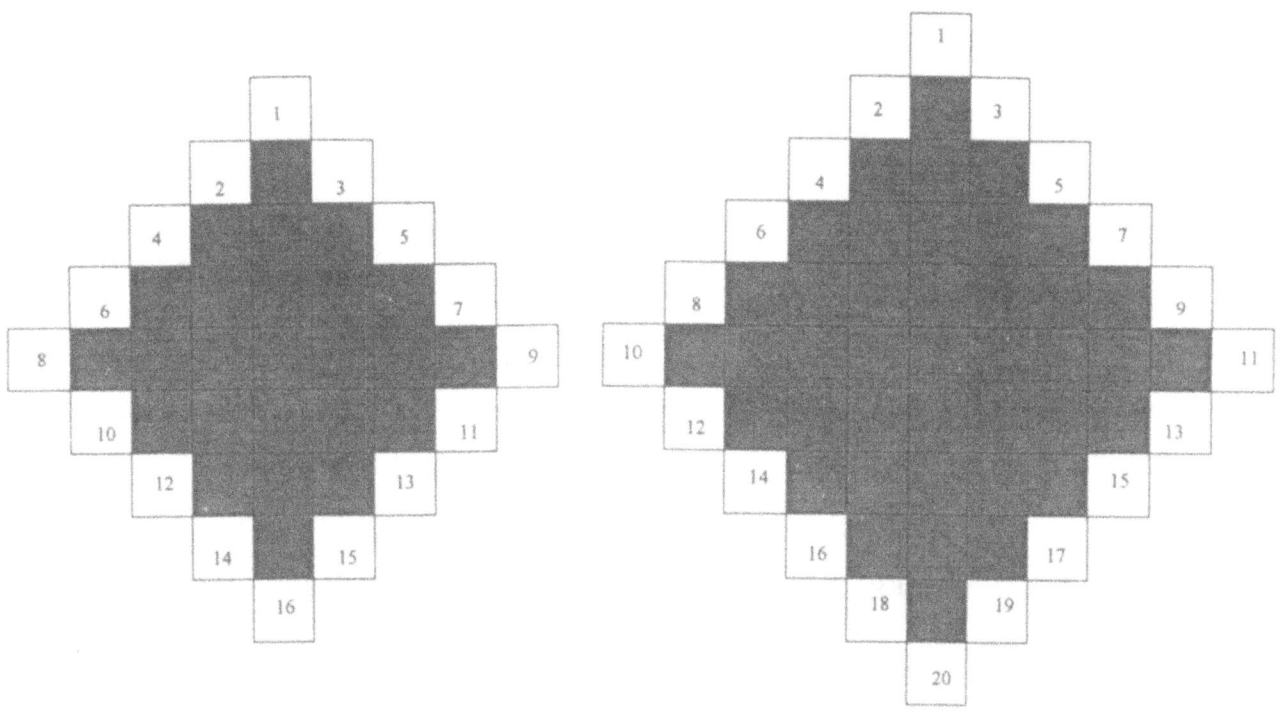

However, this is a very time-consuming process requiring a great deal of careful effort. Instead, let's examine the four figures we already have and see if there is a pattern we can use.

Figure number	Number of squares	Additional squares
1	1	
2	5	4
3	13	8
4	25	12
5	?	?
6	?	?

Although we have only four figures, the pattern of additional squares seems to be a series of numbers increasing by 4 (i.e., 4, 8, 12, 16, 20, ...). Thus we can conclude that we must add 16 squares to the fourth figure to obtain the fifth figure, and add 20 squares to the fifth to obtain the sixth. The answer to the problem is choice C.

Your students must realize that patterns play a significant role in solving problems in the *SAT I* exam, and they should learn to look automatically for patterns whenever any sequence of terms appears in a problem.

STRATEGY 3: ADOPTING A DIFFERENT POINT OF VIEW

When students are confronted by a problem that seems a bit "unusual," it may be to their advantage to consider the problem from a different point of view. While the traditional, and possibly obvious, approach may eventually yield the correct answer, you should point out that, on an examination such as the *SAT I* where time is a critical factor, a different point of view may well yield the answer in a more efficient manner. It is a good idea to practice this by examining all problems in as many different ways as possible.

ILLUSTRATIVE PROBLEM

Find the value of $x + y$, if $x^2 + y^2 = 36$ and $xy = -10$.

(A) ±26   (B) ±16   (C) ±8   (D) ±4   (E) None of these

ANALYSIS

The initial reaction of most students to this problem is to attempt to solve the two given equations simultaneously. After all, they are given two equations and two variables; the simultaneous solution should lead to values for x and y. However, this proves to be a rather tedious process. Furthermore, you should point out to them that neither equation is linear—one has a graph of a circle and the other yields a rectangular hyperbola. There might be as many as four sets of answers. Let's take a different point of view for a moment. Students should ask themselves why the problem is calling for "the value of $x + y$" and *not* the individual values of x and y. How are $x^2 + y^2$ and xy related? Where have they seen these two expressions before? Aha! Both appear in the expansion of $(x + y)^2$. So,

$$(x + y)^2 = x^2 + y^2 + 2xy.$$

But $x^2 + y^2 = 36$ and $xy = -10$. Thus,

$$(x + y)^2 = x^2 + y^2 + 2xy = 36 + (-20) = 16$$
$$(x + y)^2 = 16$$
$$x + y = \pm 4.$$

The correct answer is (D) ±4, easily arrived at once we looked at the problem from a different point of view.

ILLUSTRATIVE PROBLEM

If it is now 10:45 a.m., what time will it be 143,999,999,995 minutes later?

(A) 10:40 a.m.   (B) 10:40 p.m.   (C) 10:45 a.m.
(D) 10:45 p.m.   (E) None of these

ANALYSIS

Again, the problem appears to be rather complex. While a large number such as 143,999,999,995 suggests the possible use of a calculator, this is a 12-digit numeral, hardly likely to "fit" in the calculator display. Let's examine this problem from another point of view. The number 143,999,999,995 is only five minutes short of being 144,000,000,000 minutes, an exact multiple of 60. But 144,000,000,000/60 = 2,400,000,000 hours = 100,000,000 days. In this case, the time would also be 10:45 a.m. Since the given number was 5 minutes short of that, the correct time would be 10:40 a.m., choice A. Again, take this opportunity to point out to your students the obvious advantage of examining the problem from a different point of view.

Illustrative Problem

A rectangle has adjacent sides of 10 units and 8 units, respectively. A second rectangle with adjacent sides of 6 units and 4 units overlaps the first rectangle as shown in the figure. What is the difference in area between the two non-overlapping regions of the two rectangles?

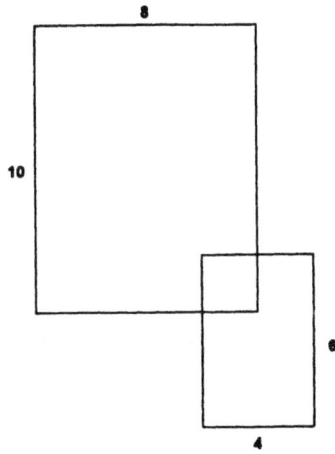

(A) 80 square units   (B) 56 square units   (C) 24 square units
(D) Cannot be determined with given information
(E) None of the above

Analysis

If students attempt to attack this problem directly, they will quickly find themselves at a loss. However, let's consider the drawing from a different point of view; that is, since the region of overlap is not determined or specified, let us suppose the rectangles are attached only at a single corner, with a region of area 0 as overlap.

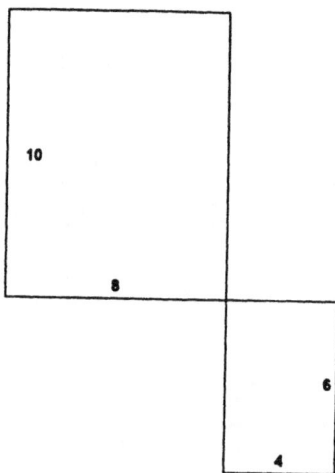

Now we see that the difference is 80 square units - 24 square units = 56 square units. Next consider the rectangles so that there will be a single unit of overlap.

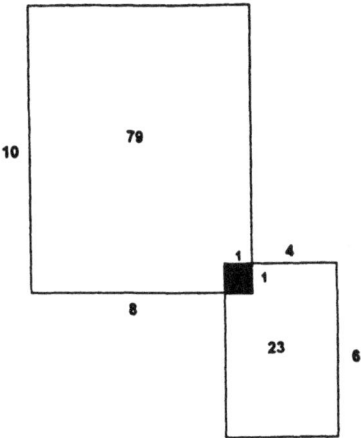

The non-overlapping difference is 79 square units - 23 square units = 56 square units. Finally, let's consider the smaller rectangle as entirely enclosed within the large one.

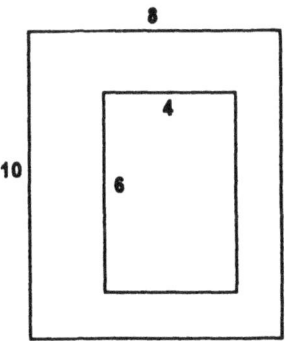

Again the non-overlapping region is 56 square units. The correct answer is (B) 56 square units. Once again, emphasize to your students the value of considering a problem from a different point of view.

Notice that you need not consider each of the situations as described above. For some students, the initial case of the rectangles merely sharing a common point might be sufficient. For others, however, the other situations might be necessary for them to conclude the correct answer.

In the previous examples, what originally appeared to be complex problems requiring extensive calculations became much easier to solve when the students examined the problems from a different point of view. You should take every opportunity to emphasize this technique to your students and encourage them to examine all problems from more than one obvious point of view. The next problem illustrates this quite vividly.

ILLUSTRATIVE PROBLEM

What is the smallest prime number that divides the sum $5^7 + 7^{11} + 11^{13} + 13^5$?

(A) 2   (B) 3   (C) 5   (D) 11   (E) None of these

ANALYSIS

At first glance, your students may decide to attempt to find the actual value for each term—5^7, 7^{11}, 11^{13}, and 13^5. Then they hope to find the sum of these four terms and test the various prime divisors. This will take a great deal of time (even if they do it correctly!). There must be another way to approach this problem. Since no factors of 2 are included, 5^7 is odd, 7^{11} is odd, 11^{13} is odd, and 13^5 is odd. And since the sum of four odd numbers must be even, the smallest prime number that divides the sum will be 2, choice A. The problem was easily resolved by examining it from an alternate point of view.

ILLUSTRATIVE PROBLEM

In the given figure, three rectangles, ABCD, BGFE, and GHIJ, are placed together as shown, and a straight line is drawn connecting point A to point I. The dimensions of rectangle ABCD are AD = 1, AB = 2. The dimensions of rectangle BGFE are BE = 2, BG = 4. The dimensions of rectangle GHIJ are GJ = 4, GH = 8. What is the area of the shaded region?

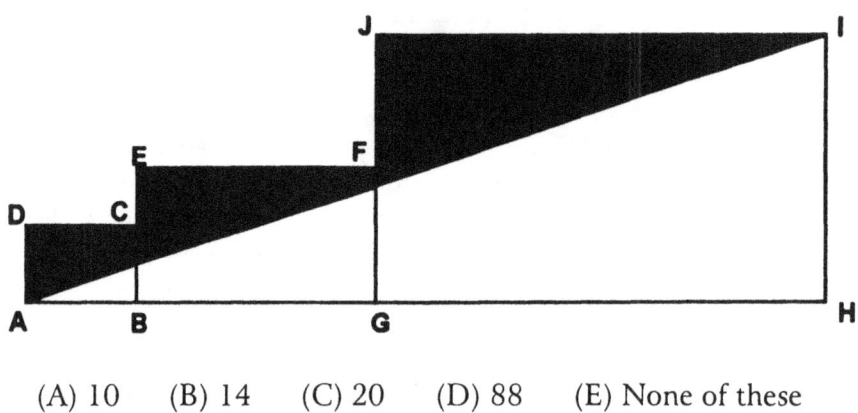

(A) 10   (B) 14   (C) 20   (D) 88   (E) None of these

ANALYSIS

The figure whose area we seek is rather oddly shaped. Students will first try to find the area of each of the shaded triangles. However, this should prove impossible, since the location of the points where $\overline{AI}$ crosses each of the sides of the rectangles is unknown. Let's approach this problem from a different point of view. We can find the area of the three rectangles and then subtract the area of the *unshaded* portion. The unshaded portion is simply a right triangle whose legs are 14 and 4. The area of the three rectangles is quite readily found:

$$\begin{aligned} \text{Rectangle ABCD} &= 2 \cdot 1 = 2 \\ \text{Rectangle BGFE} &= 4 \cdot 2 = 8 \\ \text{Rectangle GHIJ} &= 8 \cdot 4 = 32 \\ \text{Total area} & = 42 \end{aligned}$$

The area of the right triangle AIH = $(14 \cdot 4)/2 = 28$. The area of the shaded portion is $42 - 28$, or 14. The correct answer is choice B.

Adopting a different point of view can often take a seemingly complex problem from the realm of the time consuming and quite difficult into the realm of the relatively simple. Students must learn to consider approaching problems from a point of view that may be somewhat different from the very obvious one that occurs to them almost immediately.

ILLUSTRATIVE PROBLEM

Find the value of the following:

$$20 - 19 + 18 - 17 + 16 - 15 + 14 - 13 + 12 - 11$$

(A) 8   (B) 5   (C) 3   (D) 11   (E) None of these

ANALYSIS

Obviously, the students can punch the sequence of numbers into their calculators and arrive at the answer. However, this will take time to do, and errors are quite possible. If the student examines the problem from a slightly different point of view, he or she may be able to get the result quicker. Notice that each pair of numbers can be grouped as follows:

$$(20 - 19) + (18 - 17) + (16 - 15) + (14 - 13) + (12 - 11)$$
$$1 \; + \; 1 \; + \; 1 \; + \; 1 \; + \; 1$$

which quickly yields the correct answer: (B) 5.

Illustrative Problem

In the given figure, ABCD and BDFE are parallelograms. What is the ratio of their areas?

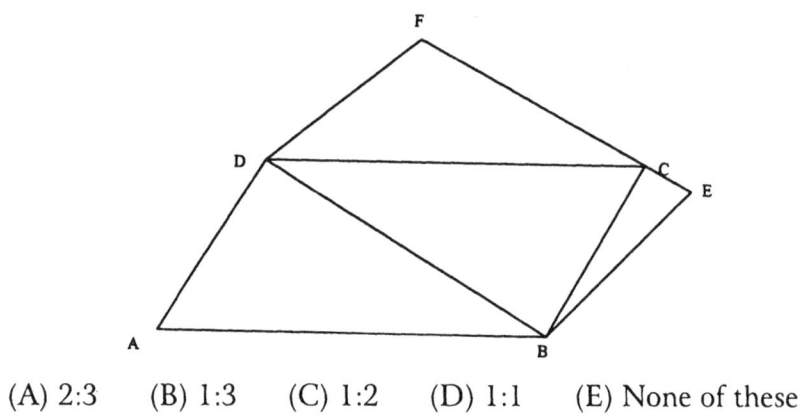

(A) 2:3   (B) 1:3   (C) 1:2   (D) 1:1   (E) None of these

Analysis

Students' immediate reaction is to examine each of the parallelograms and try to find their individual areas. However, let's examine the problem from a different point of view. In parallelogram ABCD, △BCD = ½ the area of ABCD, since $\overline{BD}$ is a diagonal. In parallelogram BDEF, △BCD = ½ the area of BDFE, since they have a common base ($\overline{BD}$) and the same altitude (a perpendicular from C to $\overline{BD}$). Thus the remaining parts of the two parallelograms also have equal areas, and the ratio of the areas of the two parallelograms is 1:1, choice D.

Illustrative Problem

Find the difference between the areas of the two right triangles ABC and BCD, where BC = 18, BD = 30, and AD = 9.

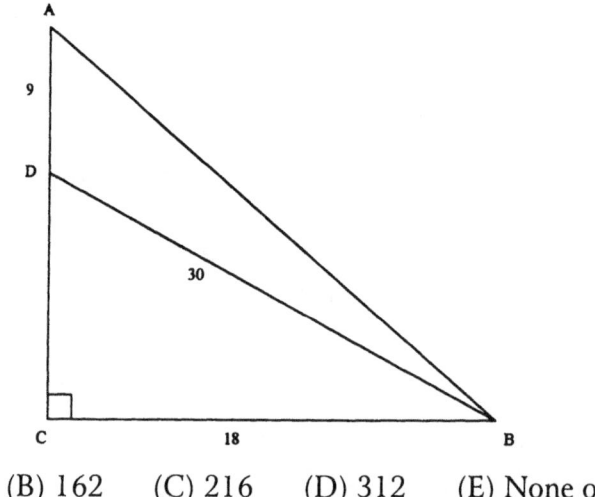

(A) 81   (B) 162   (C) 216   (D) 312   (E) None of these

ANALYSIS

We can find the areas of the two right triangles in question and then take the difference between them. However, this would prove to be quite time consuming, for we would first have to use the Pythagorean theorem to find DC = 24. Perhaps there is another way to solve the problem by considering it from a different point of view.

If we revisit the question that asks for "the difference between the areas," this is, geometrically speaking, △ADB. This obtuse triangle, with base AD = 9 and altitude BC = 18, has an area computed directly as A = ½(9)(18) = 81 square units, or choice A.

ILLUSTRATIVE PROBLEM

The four points A (3,3), B (5,7), C (8,7), and D (12,3) are the vertices of a trapezoid. The points A′, B′, C′, and D′ are found by multiplying each of the abscissas of A, B, C, and D, respectively, by -1. Find the difference in the number of square units in the areas of trapezoid ABCD and trapezoid A′B′C′D′.

(A) 15   (B) 6   (C) 4   (D) 0   (E) None of these

ANALYSIS

Obviously, students can resolve the problem by finding the area of each trapezoid and then finding the difference. Point out to them, however, that we can solve this problem easily by looking at it from a different perspective. Examine the operator X → X′. Notice that this results in a translation of the original figure, and effects no change in area. The correct answer is (D) 0. Notice that multiplication of the abscissas by -1 merely moves the vertices of the trapezoid to the opposite side of the y-axis. It does nothing to the size or shape of the original figure.

ILLUSTRATIVE PROBLEM

In the figure below, ABC is an equilateral triangle with KLMNPQ a regular hexagon. What is the ratio of the area of the hexagon KLMNPQ to the area of △ABC, if KL = ⅓(AB)?

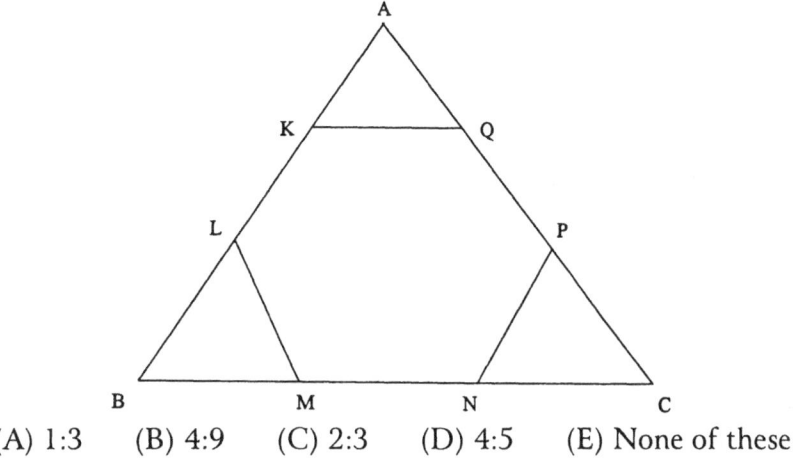

(A) 1:3   (B) 4:9   (C) 2:3   (D) 4:5   (E) None of these

ANALYSIS

While we could assign a value to the side of the equilateral triangle and then find a side of the regular hexagon, we would still have to find the area of each figure. It might be easier and quicker to adopt a different point of view—namely, to draw a series of lines as shown in the figure, dividing the original triangle into 9 congruent, equilateral triangles.

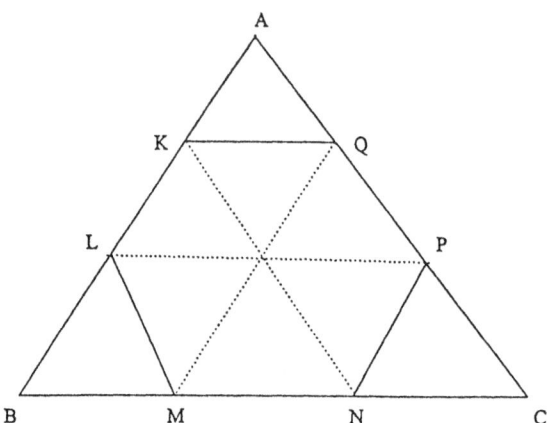

The hexagon KLMNPQ comprises 6 of these, whereas the original triangle comprises 9. Thus the ratio of their areas would be 6:9, or (C) 2:3.

ILLUSTRATIVE PROBLEM

You have a nickel, a dime, a quarter, and a half-dollar. A clerk shows you several items, each of which has a different price and any one of which you can buy with one or more of your coins, without receiving any change. What is the maximum number of items the clerk could have shown you?

(A) 8   (B) 10   (C) 13   (D) 15   (E) None of these

ANALYSIS

The number of coins to be used to pay for the article is not specified—you could use 1 or 2 or 3 or all 4. Using only 1 coin, there are 4 possibilities: N, D, Q, H. Using any 2 of the coins, there are 6 possibilities: ND, NQ, NH, DQ, DH, QH. Using any 3 coins, there are 4 possibilities: NDQ, NDH, NQH, DQH. Using all 4 coins gives only 1 possibility. Thus, there are a total of 15 different possibilities. However, the problem can easily be resolved by adopting a different point of view, and this should be pointed out to your students. Each coin has exactly 2 possibilities—it can be spent or not spent. Thus, there are $2^4 = 16$ possibilities. But this includes the one situation in which you spend 0 coins (impossible), so the final answer is 16 - 1 = 15, choice D.

Illustrative Problem

In the figure shown, the ratios of the diameters of the three circles is 2:3:5. Find the ratio of the area of the shaded region to the area of the largest circle.

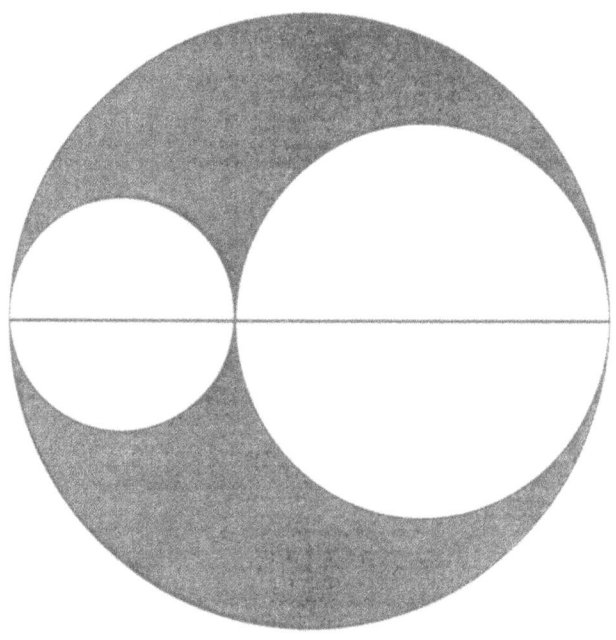

(A) 2:3   (B) 12:25   (C) 13:25   (D) 2:5   (E) None of these

Analysis

Since the ratio of the linear parts (the diameters) is given as 2:3:5, the respective areas must be in the ratios of $2^2:3^2:5^2 = 4:9:25$. Let's assume that the area of the largest circle is 25. Then the areas of the other two circles are 4 and 9, for a sum of 13. The area of the shaded region would be 25 - 13 = 12. The required ratio is 12:25, choice B.

Illustrative Problem

Given rectangle ABCD with AB = 12, BC = 9, EB = BF = 3, find the area of △DEF.

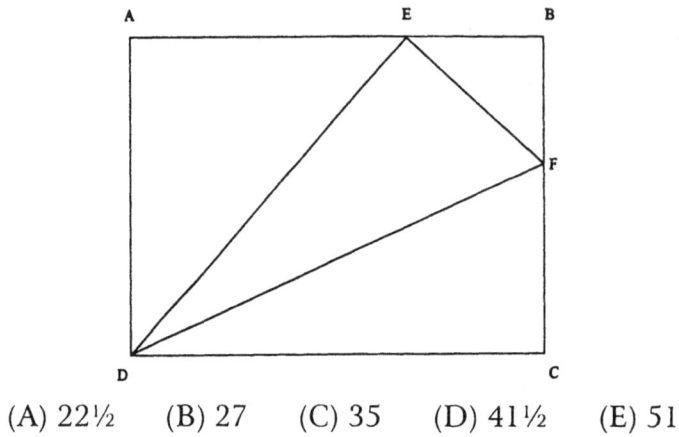

(A) 22½   (B) 27   (C) 35   (D) 41½   (E) 51

Analysis

Students may try to find the area of △DEF directly. This is a most difficult task. So, let's approach the problem from another point of view. We'll find the areas of triangles ADE, BEF, and DFC and subtract their sum from the area of the rectangle.

Area △AED = ½(9 · 9) = ½(81)
Area △BEF = ½(3 · 3) = ½(9)
Area △FCD = ½(6 · 12) = ½(72)
Total area = ½(162)   = 81

The entire rectangle has area = 12 · 9 = 108. Thus the area of △DEF = 108 - 81 = 27, choice B.

Strategy 4: Solve A Simpler, Analogous Problem

At first reading, some problems appear extremely complex. In some cases, this may discourage your students from even attempting a problem. Sometimes it is the magnitude of the numbers that makes the problem appear difficult; other times it may be the excessive amount of data given in the problem, or even the way in which the problem has been stated. Regardless of the reason, before deciding that a problem is too difficult or complex to attempt, students should try converting the problem into a simpler but equivalent version of the original problem. Show your students how to change the numbers into simpler ones, how to consider a modified drawing of the problem, or some other way to approach the problem by putting it in a simpler form.

Solving this simpler version of the problem may enable the students to see which approach can then be taken to solve the original problem. Only practice can help students develop this problem-solving skill.

ILLUSTRATIVE PROBLEM

Find all the integral values of x that satisfy $(3x + 7)^{x^2 - 9} = 1$

(A) 3,4   (B) 3,-2   (C) -2,5   (D) 1,2   (E) None of these

ANALYSIS

The sight of a linear expression raised to a quadratic exponent gives the student the feeling that a rather lengthy and complex solution is required. Many students will not even attempt the problem, considering the amount of time required to be inordinate for the credit they will receive. Consequently, these students may decide to omit the problem and move on. Others may decide to try the various answers given, substituting them into the original problem. Point out to these students that this, too, is very time consuming and may not work, since all the real values of x may not have been given.

However, let's examine a simpler, analogous version of the problem to find out what is really taking place here. For example, let's look at $a^b = 1$. This problem is a bit less frightening, and somewhat easier to examine and discuss. The expression will have a value of 1 when the base, a, is 1, since $(1)^b = 1$ for any value of b. Similarly, the expression will have value 1 when the exponent is 0, since $(a)^0 = 1$ for any nonzero value of a. The expression will also have a value of 1 when $a = -1$ and b is even. Now we have a method for attacking the original problem. Have students go back to the original problem and apply what they have found in their examination of the simpler version.

Case I
   Since 1 raised to any power = 1, set the base = 1.
   $3x + 7 = 1$
   $3x = -6$
   $x = -2$

Case II
   Since any nonzero expression raised to the 0 power = 1, set the exponent = 0.
   $x^2 - 9 = 0$
   $(x - 3)(x + 3) = 0$
   $x = 3$ $x = -3$

Case III
   When -1 is raised to an even power, it also has a value of +1. Consider $3x + 7 = -1$, then $x = -8/3$, which is nonintegral.

Thus, there are 3 integral values of x for which the equation is correct: +3, -3, and -2. Choice E is the correct answer.

Illustrative Problem

What is the units digit of 7^{23}?

(A) 9   (B) 7   (C) 3   (D) 1   (E) None of these

Analysis

This problem appears, on the surface, to be rather complex. If your students attempt to resolve this with their calculators, they will quickly discover that the number becomes too large to fit into the calculator display. Let's reduce the problem to a simpler version of the original by taking increasingly larger powers of 7, starting with 7^1. Then we can gradually expand until a pattern emerges that we recognize and can then apply to our original problem. Notice that this problem also requires that the students make use of the pattern-recognition strategy that we have previously discussed.

$$7^1 = \underline{7} \qquad 7^5 = 1690\underline{7}$$
$$7^2 = 4\underline{9} \qquad 7^6 = 11764\underline{9}$$
$$7^3 = 34\underline{3} \qquad 7^7 = 82354\underline{3}$$
$$7^4 = 240\underline{1} \qquad 7^8 = 576480\underline{1}$$

Aha! The units digits in these simplified versions repeat in cycles of four. Now we can return to our original problem. Since the exponent 23 (of our original problem 7^{23}) has a remainder of 3 when divided by 4, we can assume that its terminal digit is the same as that of 7^{19}, 7^{15}, 7^{11}, 7^7, which we know is 3. The correct answer to our problem is choice C.

Once again, turning to a simpler version of the original problem has enabled us to solve the original problem in a relatively simple manner.

Now let's turn to a problem that uses variables and appears to require a great deal of computation. Again, alert your students to the need to simplify the problem, this time by replacing the variables with simple numbers.

Illustrative Problem

The basketball squad is taking part in a free-throw contest. The first player scored x free-throws. The second player scored y free throws. The third player made the same number of free-throws as the arithmetic mean of the number of free-throws made by the first two players. Each subsequent player in the contest scored the arithmetic mean of the number of free-throws made by all the players who had preceded her. How many free-throws did the twentieth player make?

(A) 20   (B) $5x + 5y$ (C) $(x + y)/2$ (D) $(x + y)/20$
(E) None of these

ANALYSIS

Some students may decide to solve this problem by finding the arithmetic mean for each player in turn, using the results of all previous players. This would take much too long! Let's examine a simplified version of the problem by replacing x and y with simple numbers. Suppose the first player made 8 free-throws and the second player made 12. Then the third player had a score equal to their arithmetic mean, or $(8 + 12)/2 = 10$. Now, the number of free-throws made by the fourth player is the arithmetic mean of the scores made by the first three players, or $(8 + 12 + 10)/3 = 10$. The fifth player's score is the arithmetic mean of the scores made by the first four players, or $(8 + 12 + 10 + 10)/4 = 10$. This reveals that the score made by any player after the first two will be the same as the arithmetic mean for the first two players' scores. We now return to the original problem. The correct answer is the arithmetic mean of the scores made by the first two players, or $(x + y)/2$, which is choice C. Once again, the simpler, analogous problem enables us to determine the method for solving the original problem rather quickly. Impress on your students the concept of using simpler numbers whenever the original numbers seem to cause confusion. Let's look at another such problem.

ILLUSTRATIVE PROBLEM

Given that m and n are each positive integers greater than 1, which of the following has the greatest value?

(A) $m + n$ (B) $\sqrt{2mn}$ (C) $(m^2 + n^2)/(m + n)$
(D) $(m^4 + n^4)/(m^3 + n^3)$ (E) $m - n$

ANALYSIS

When students are asked to "order" complicated expressions with variables in both numerator and denominator, the problem becomes much simpler if we solve an easier version, in which we substitute for the variables. Let's let $m = 2$ and $n = 4$. Then for the expression in choice A we get $2 + 4 = 6$. For choice B we get $\sqrt{16} = 4$. For C we get $(4 + 16)/(2 + 4) = 3.33\overline{3}$. For D we get $(16 + 256)/(8 + 64) = 3.77\overline{7}$. For E we get $2 - 4 = -2$. Thus, the correct answer is (A) $m + n$. Once again, solving the simplified version enabled us to solve the original problem more readily.

ILLUSTRATIVE PROBLEM

Sam's Discount Paradise offers a 30% discount on the list price of a certain article. Mitch's Lowprice Emporium usually offers a 20% discount on the same list price of the same article. However, today, Mitch's is offering an additional discount of 10% on its already reduced price. Which of the following is true?

(A) Sam's offers a 2% greater discount than Mitch's.
(B) Mitch's offers a 2% greater discount than Sam's.
(C) Both stores offer the same discount.
(D) The difference between the discounts is less than 2%.
(E) The difference between the discounts is greater than 2%.

ANALYSIS

Since most of your students are well versed in algebraic techniques, they may start the problem by letting a variable represent the price of the article and then applying the various discounts to see what is taking place. However, this will soon become rather time consuming and demand a great deal of careful computation. Let's take a look at a simpler, analogous version of the problem. Since no specific price was given for the article in question, let's assume the article was originally priced at $100. (It is often a good idea to use a base of 100 when dealing with percent problems.) Then Sam's price will be $100 - $30 = $70. Mitch's price would normally be $100 - $20 = $80. However, today we have an additional discount of 10% *based on the reduced price*. This makes the final price at Mitch's $80 - $8 = $72. Thus, Sam's price is still $2 cheaper. Since we began with the article priced at $100, the $2 discount represents a 2% greater discount at Sam's, choice A.

Students should have practice with problems dealing with "successive discounts." An alternative method for solving this kind of problem can be found on page 20 of this volume. Have your students practice problems involving both successive mark-ups (increases) and successive discounts.

ILLUSTRATIVE PROBLEM

Given 19 consecutive integers whose sum is 95, what is the 10th number?

(A) 8   (B) 5   (C) 0   (D) -5   (E) -8

Analysis

Sometimes merely recognizing that a simpler version of a particular problem will give us what we are looking for is a difficult process in itself. This problem is an excellent illustration of this. Many students will begin by writing out the 19 consecutive numbers in algebraic form: x, $(x + 1)$, $(x + 2)$, $(x + 3)$, ..., $(x + 17)$, $(x + 18)$; then they will sum these to give a sum of 95. However, there is a simpler yet analogous way to proceed. Recognizing that the 10th integer of 19 integers is the middle one, students may represent this one as x and the rest as $(x - 9)$, $(x - 8)$, $(x - 7)$, ..., $(x - 1)$, x, $(x + 1)$, ..., $(x + 7)$, $(x + 8)$, $(x + 9)$, and then sum these to 95. Students should notice that by pairing the first and last terms, $(x - 9) + (x + 9)$, the second and next-to-last, $(x - 8) + (x + 8)$, and so on, we arrive at $19x = 95$, $x = 5$. The correct choice is B, and we found it by looking at a simpler version of the problem.

However, we may also approach the problem in a still simpler manner. The middle term is the average (or arithmetic mean) of the given 19 integers. Thus, we simply take the sum, 95, and divide it by the number of integers, 19, and arrive at the same answer: (B) 5.

Illustrative Problem

What is the quotient of 1/500,000,000,000?

(A) .00000002   (B) .000000002   (C) .000000000002
(D) .0000000000002   (E) None of these

Analysis

The display area on most calculators that students use is too small to fit the answer. It can be done manually, although the computation often leads to an error due to the large number of 0's in the answer. Let's try a simpler version (or versions) of the problem, examine the answers we obtain, and see if a usable pattern emerges. Notice that, once again, the recognizing of a pattern strategy becomes an integral part of another strategy as well.

	Number of 0's after the 5	Quotient	Number of 0's between the decimal and the 2
1/5	0	.2	0
1/50	1	.02	1
1/500	2	.002	2
1/5000	3	.0002	3
⋮	⋮	⋮	⋮
1/500000000000	11	.000000000002	11

The correct answer is now easily found. The number of 0's after the decimal point and before the 2 is the same as the number of 0's in the divisor. The correct answer is choice C.

Strategy 5: Using Extreme Cases

Some problems can be much more easily solved by considering extreme cases of the situation under consideration. By considering extremes we are changing variables in the problem, but those that do not affect the actual problem situation. In everyday life this method of problem solving can be seen in a question such as "Will I stay drier by running or walking in a rainstorm?" One's first reaction may be to run, but then further thought may lead to uncertainty when one remembers that the accumulation of water on the windshield of a car appears to increase with an increase in speed. We can even analyze the wetness of the various parts of the body (e.g., the top and front) in the rainstorm and the effects of running or walking in the rainstorm (such an analysis can be found in *The Art of Problem Solving: A Resource for the Mathematics Teacher* by Posamentier and Schulz [Thousand Oaks, CA: Corwin Press, 1996], p. 92). We can attempt to better understand the problem as a "real life" problem by simply asking, "What happens if we were to go *extremely* slowly in the rain?" We would get drenched! The answer to the question is then obvious: We would be best off minimizing the time we are exposed to the rainstorm. By considering the extreme case, we were able to put the problem in a more solvable light. This strategy is clearly not applicable to every problem-solving situation, but then not every strategy can be used for all problems. After pointing out to students when certain strategies can be applied, they will begin to develop a sensitivity to recognizing when to use specific methods. An interesting mathematical problem will help shed further light on this method.

Illustrative Problem

Two concentric circles are 10 units apart. What is the difference between the circumferences of the circles?

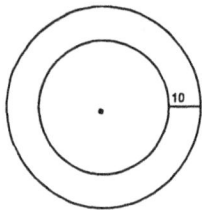

(A) 5π (B) 10π (C) 15π (D) 20π
(E) Cannot be determined from information given

Analysis

The traditional method of solving this problem is to find the radii of the two circles and then take the difference in the circumferences of the two circles. Since we do not know the circumferences, it appears that we cannot find their difference. A clever approach would be to consider an extreme case.

Suppose that the smaller of the two circles gets smaller and smaller until it reaches the extreme case and becomes a point: the center of the larger circle. The distance between the circles now becomes the radius of the larger circle, and the "difference between the circumferences" is actually the circumference of the larger circle. Hence the sought-after difference is $2 \cdot \pi \cdot 10 = 20\pi$.

As another example of how this method of problem solving can be effective, consider the following.

ILLUSTRATIVE PROBLEM

If it takes 5 people 3 days to paint the gymnasium of a school, how many days will it take 7 people working at the same rate?

(A) 15   (B) 7   (C) $4\frac{1}{5}$ (D) $2\frac{1}{7}$ (E) None of these

ANALYSIS

Although in this book we consider various ways to solve this sort of problem, one method can involve using the extreme case. Suppose, in the extreme case, only one person is painting the gymnasium; how many days would it take to do the entire job? Five times as long, or 15 days. Now, to find the time required for 7 people to do the same job (of course, all the while assuming that all people work at the same rate) we simply divide by 7 to get $\frac{15}{7} = 2\frac{1}{7}$, or ≈ 2.14 days. This illustration of the technique, although quite common, deserves special attention by the teacher so that students can become accustomed to employing this method when appropriate. A more essential application of the method of using extreme cases to solve problems can be seen in the next example.

ILLUSTRATIVE PROBLEM

What is the sum of the measures of the vertex angles (marked) of any pentagram? (See figure below.)

(A) 90   (B) 180   (C) 360   (D) 540   (E) 720

ANALYSIS

The problem is solvable using an irregular pentagram, which would make the solution follow from the given diagram. However, since the problem did not specify the type of pentagram, it calls for an answer that would hold true for any pentagram. This gives us license to use an extreme case: either one that can be inscribed in a circle or one that is regular (i.e., all the vertex angles are the same measure).

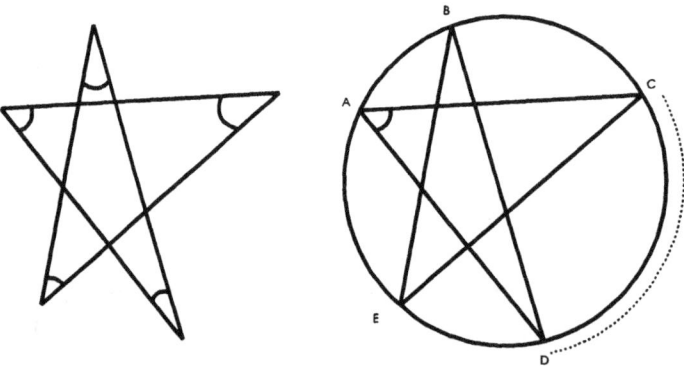

In the case where the pentagram is inscribable in a circle, each angle is the inscribed angle of a distinct arc of the circle. Each angle measures, therefore, one-half the measure of one of the arcs of the circle. And all arcs determined by the vertices are used exactly once. Therefore, the sum of the vertex angles is one-half the sum of the arcs, or ½(360°) = 180°.

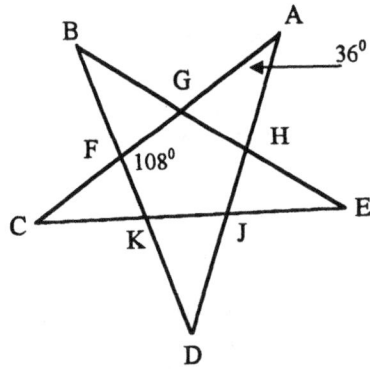

In the extreme case where the pentagram ABCDE is regular, pentagon FGHJK is also regular, with each angle 108°; thus for isosceles triangle AFD, whose vertex angle is 108°, each base angle must be 36°. Again it must be noted that we considered an extreme case here since nothing of the given information was compromised. This is an important notion to impart to students impressed with this nice strategy for rather simply solving seemingly difficult problems. The answer is (B) 180.

Strategy 6: Visual Representation

In many problems, students are presented with a great deal of data, usually in verbal form. In order to see "what is going on," students must organize their data in some visual format—make a diagram of the situation, prepare a table of the data, create a chart from the information, and so on. You must stress the necessity for carefully organizing the data in a problem and the importance of labeling the diagram, table, or chart as clearly as possible as the students develop it, or else confusion may arise.

Visual representation often reveals a situation that the students are already quite familiar with and that would not have been as obvious if the problem remained in verbal form only.

Not all problems are obvious in their call for a visual representation. In some cases, however, the drawing is very useful in order to see what is actually taking place.

Illustrative Problem

It takes 7 seconds for a clock to chime 7 times at 7:00. At this same rate, how many seconds will it take the clock to chime at 10:00? (Assume that each actual chime takes no time at all.)

(A) 9   (B) 9.5   (C) 10   (D) 10.5   (E) 11

Analysis

To determine the action in this problem, let's draw a diagram of the clock striking 7:00. In our diagram, each line segment represents a chime by the clock.

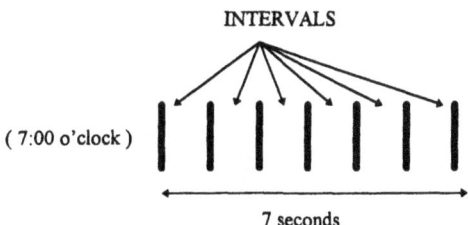

We see that there are 6 intervals. Thus, each interval is 7 ÷ 6, or 7/6 seconds long. Now let's see what happens when the clock strikes 10:00.

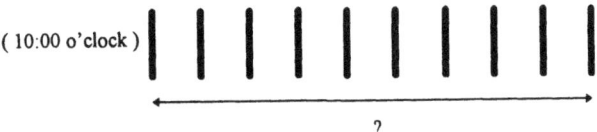

Now we see that there are 9 intervals. Since we know that each interval is 7/6 seconds long, it takes the clock 9 · (7/6) = 63/6, or 10.5 seconds, choice D.

Illustrative Problem

Fly-by-Night Airlines has a daily flight from New Orleans to New York at 6:16 p.m. On last Tuesday's flight there were 480 passengers aboard. Of these, 1/8 flew in first class, while the rest flew in coach class. Of those in coach, 2/3 were members of the City College Alumni Association, on the way to a reunion. Of the Alumni Association members on board and flying in coach, 1/4 were female. How many female passengers in coach class were from the Alumni Association?

(A) 60   (B) 70   (C) 140   (D) 280   (E) 420

Analysis

There is a great deal of information given in this problem. In order better to grasp the situation, students may feel it necessary to make a visual representation of the situation. Let's represent the entire passenger population with a large rectangle.

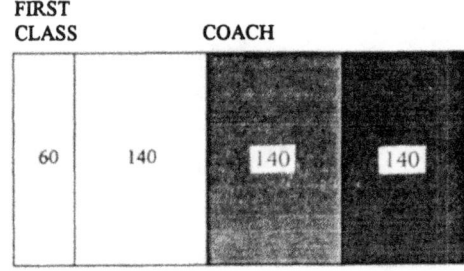

Now we can divide the coach portion into thirds, and take 2 of them:

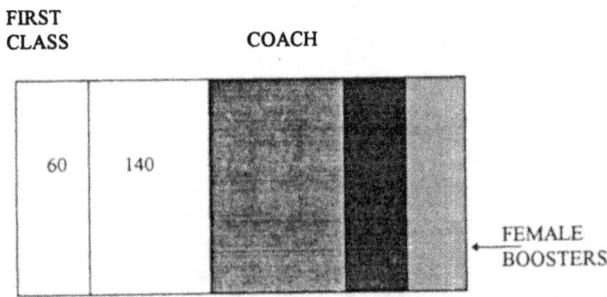

We finally take 1/4 of these two thirds:

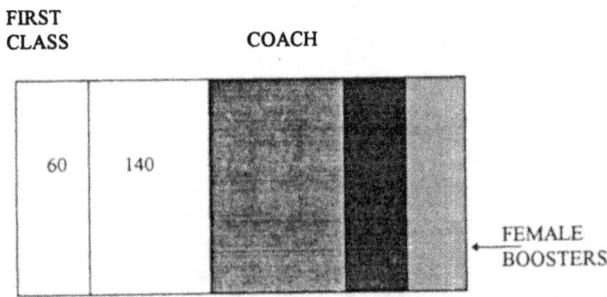

At this point, the answer is easily found to be ½ of 140, or 70, which is choice B.

ILLUSTRATIVE PROBLEM

When the walls of a typical house are put into place, they are supported by a series of 2 × 4-inch studs from end to end, onto which the wall is nailed. These studs are placed 16 inches apart, from the center of one to the center of the next one. How many studs would be needed for a wall 48 feet long?

(A) 28   (B) 36   (C) 37   (D) 42   (E) None of these

ANALYSIS

The students' initial reaction is simply to divide 48 feet by 16 inches. This will give us 36. However, if we try to visually examine what is taking place, we see that this fails to account for the fact that there must be a stud at each end.

Thus, the correct answer is (C) 37.

ILLUSTRATIVE PROBLEM

Amanda left her house to drive to City Hall in Boise. Her car's odometer read 32,518 miles when she started. After she had gone 5 miles, she realized that she had left her briefcase at home. She returned home, picked up her briefcase, and drove to City Hall. After her meeting, she turned around and drove directly home. Her odometer now read 32,966. How many miles from her home is City Hall?

(A) 219   (B) 229   (C) 448   (D) 528   (E) None of these

ANALYSIS

This is obviously not the typical (or expected) "uniform motion" problem. However, in order to better visualize the given information in the problem, your students should be encouraged to draw a visual representation of the situation and see exactly what is taking place.

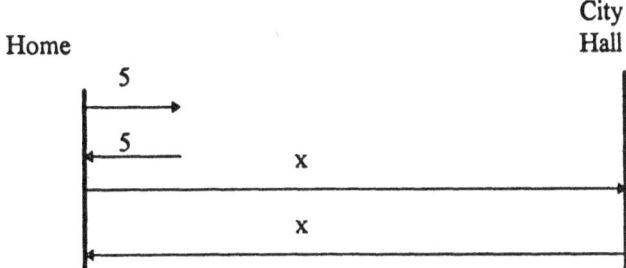

If we let x = the distance from Amanda's house to Boise City Hall, we obtain the simple equation:

$$2x + 10 = 32{,}966 - 32{,}518$$
$$2x + 10 = 448$$
$$2x = 438$$
$$x = 219$$

The correct answer is choice A.

ILLUSTRATIVE PROBLEM

Two tests contain 25 different questions each. If the first 5 questions on test I are added to the end of test II, and the first 5 questions from test II are added to the end of test I, each test now has 30 questions. How many questions now are the same on both tests?

(A) 5   (B) 10   (C) 15   (D) 20   (E) 30

ANALYSIS

Let's chart a before-and-after visual representation of what has taken place in this problem:

Before	Test I	A	B	C	D	...	W	X	Y					
	Test II	1	2	3	4	...	23	24	25					
After	Test I	A	B	C	D	E	...	W	X	Y	1	2	3	4	5
	Test II	1	2	3	4	5	...	23	24	25	A	B	C	D	E

The tests now contain 10 questions in common (i.e., 1, 2, 3, 4, 5 and A, B, C, D, E). The correct choice is B.

ILLUSTRATIVE PROBLEM

There are 12 trains that pass through towns P and Q. Of these, 1/2 stop at P, 1/3 stop at Q, and 1/4 stop at both P and Q. How many stop at neither?

(A) 5   (B) 6   (C) 7   (D) 11   (E) None of these

ANALYSIS

Let's solve this problem with a Venn diagram. Let the rectangle represent the 12 trains.

$(1/2)12 = 6$ stop at P
$(1/3)12 = 4$ stop at Q
$(1/4)12 = 3$ stop at both P and Q

The circles representing P and Q intersect, and this intersection represents those trains that stop at both P and Q.

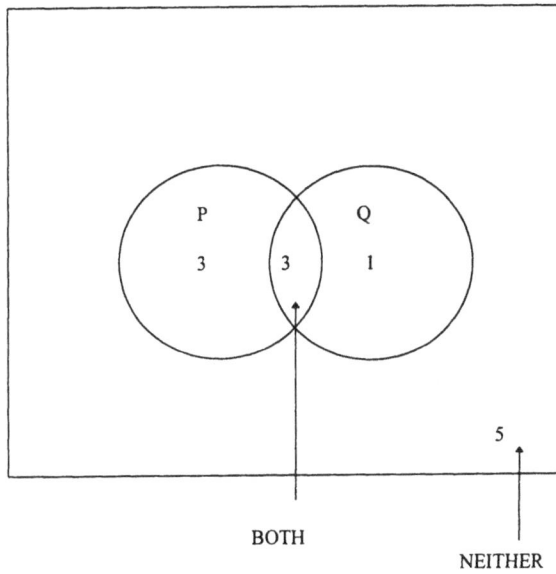

So we place the 3 in this section first. Since town P has 6 trains stopping, the rest of circle P must contain 3. Similarly, the rest of circle Q contains 1. Thus there are 5 that stop at neither, choice A.

STRATEGY 7: INTELLIGENT GUESSING AND TESTING

Many problems can be solved by making use of a guess-and-test strategy. Guesses are not made in a vacuum, nor should they be simply wild guesses, coming out of nowhere—hence this strategy takes the adjective *intelligent*. Encourage your students to make reasonable, intelligent guesses. Students test these guesses in terms of the information found in the problem, and then revise each guess based on the results of the testing process. Then they make another guess, and continue in this manner until the correct answer has been found. Students may discover, for example, that some problems that can be resolved by solving a pair of equations simultaneously in two variables can just as easily be resolved by intelligent guessing and testing. Encourage your students to practice solving problems in both ways. This should help them gain speed and accuracy in using the guess-and-test strategy. Have them explain the basis of their original and subsequent guesses.

ILLUSTRATIVE PROBLEM

Find the next two terms in the following sequence:

1, 0, 2, 3, 3, 8, 4, 15, 5, ___, ___

(A) 24, 6   (B) 6, 24   (C) 30, 7   (D) 7, 30   (E) None of these

ANALYSIS

We have a series that appears to have terms going up and down in some random order. This might suggest that there are possibly two interwoven sequences in the set of numbers. Using this intelligent guess, let's examine the sequence further:

Sequence #1	1	2	3	4	5	(the odd-place terms)
Sequence #2	0	3	8	15		(the even-place terms)

Our guess was correct; there are indeed two interwoven sequences. The first sequence is simply whole numbers ascending by ones. The second sequence is a bit more complex—the difference between successive terms is 3, 5, 7. The next term in sequence #2 will differ from 15 by 9, so it is 24. The next term in sequence #1 is 6. The correct answer, then, is (A) 24, 6. An alternative strategy would be to find a pattern—students should realize that more than one method can be applied to most problems.

ILLUSTRATIVE PROBLEM

Bill took a 20-question multiple-choice test last Thursday. The test was scored +5 for a correct answer, -2 for an incorrect answer, and 0 if the question was not answered. Bill scored 44, even though he omitted some of the questions. How many questions did he omit?

(A) 10   (B) 7   (C) 5   (D) 2   (E) None of these is correct

ANALYSIS

Students usually attempt to solve this problem by setting up a series of equations. However, this proves difficult, since we have three variables and only two equations. This will require some careful refining on the part of the students, as they substitute values into the equations. Clearly, there must be another method—in this case, intelligent guessing and testing. We first reason that Bill must have had at least 10 questions correct, or he could not have reached a score of 50 or more. If we begin with a guess of 10 correct, we can quickly see that he had 3 wrong (10 correct = +50, 3 wrong = -6, for a total score of 44). Thus, there were 7 questions left unanswered. But is there another answer? What if he had 11 correct? Impossible: Since $11 \cdot 5 = +55$, we could not arrive at a score of 44. What about 12 correct? Then he could have had 8 wrong ($12 \cdot 5 = +60$; $8 \cdot -2 = -16$). But this would mean he had not omitted any questions. The only possible answer is (B) 7.

ILLUSTRATIVE PROBLEM

The king and queen have a custom of giving gold coins to their children on their wedding anniversary. Last year, they gave each of their sons as many gold coins as they had sons, and they gave each daughter as many gold coins as they had daughters. Altogether they distributed 841 gold coins. If you were asked how many sons and how many daughters the king and queen have, how many sets of answers are there to this problem?

(A) 1 (B) 2 (C) 4 (D) 6 (E) 8

ANALYSIS

Let the number of sons be x and the number of daughters y. We obtain the equation $x^2 + y^2 = 841$. (Notice that each son receiving as many coins as there are sons gives us $x \cdot x$, or x^2). Some of your students may recognize this equation as yielding the graph of a circle whose radius is $\sqrt{841}$, or $r = 29$. They may decide to sketch the graph and consider all lattice points in the first quadrant only (since the number of sons and the number of daughters cannot be negative). This approach is neither accurate nor easily done. It would make better sense to use intelligent guessing and testing.

We can reason that there must be fewer than 30 sons/daughters, since 30 coins for each of 30 sons would equal 900, which is greater than our 841. Thus, we can begin with a guess of $x = 29$ and continue examining integral values for x to obtain those integral values of x for which we will get integral values for y. A calculator is a great help in this task. Notice that the symmetry of the problem tells us that whatever is true about the variable x is also true for the variable y. By following this procedure of trying the decreasing integral values for x from 29 to 0, we obtain the following possible answers to the problem:

Daughters (y)	0	20	21	29
Sons (x)	29	21	20	0

Some of your students may argue that the pairs of the answer with 0 sons and 29 daughters or 0 daughters and 29 sons are inappropriate, since the problem seems to indicate that there must have been both sons and daughters. However, this is not so—the problem states only what is given to sons and to daughters. They still might have had all sons or all daughters. Thus, there are 4 sets of answers to the problem—choice C. Intelligent guessing and testing enabled us to find the four sets quickly.

ILLUSTRATIVE PROBLEM

Mrs. Chamberlain gave her son $4.05 in dimes and quarters. She gave him 5 more quarters than dimes. How many dimes did Mrs. Chamberlain give her son?

(A) 5   (B) 7   (C) 8   (D) 10   (E) None of these

ANALYSIS

We could create a complete table of all the possible combinations of dimes and quarters. Or we might set up a system of two equations in two variables and solve them simultaneously:

$$q = d + 5$$
$$25q + 10d = 405$$

However, both of these processes would be rather time consuming, even assuming that your students can set up the equations. Let's use the strategy of intelligent guessing and testing. From the choices given, pick a number of dimes—say, 5. Then this would give us 10 quarters, since there must be 5 more quarters than dimes.

$$5(.10) + 10(.25) = 3.00$$

The amount is too small! Let's increase the number of dimes—try 8. We then have $8 + 5 = 13$ quarters.

$$8(.10) + 13(.25) = 4.05$$

This gives us the correct answer, 8 dimes, which was choice C.

It is a good idea to have students practice more than one procedure in solving a problem. Even if your students can use the algebraic solution, have them practice the guess-and-test strategy as well, since both can be advantageous in different situations.

ILLUSTRATIVE PROBLEM

Jean and Miguel bought a rectangular rug in Turkey last summer. The rug has an area of 40.5 square feet. The length of the rug is twice the width. What is the width of the rug?

(A) 2.5'   (B) 5'   (C) 4.5'   (D) 9'   (E) None of these

ANALYSIS

Although your students may be able to solve this problem algebraically, the solution will involve a quadratic and a linear equation:

$$lw = 40.5$$
$$l = 2w$$

A more effective approach would be to use intelligent guessing and testing. Make a series of guesses for the width of the rug, double each width to get each corresponding length, and then check the area.

Width	Length	Area	
10	20	200	(too large)
5	10	50	(still too large)
4	8	32	(too small)

The width of the rug must lie somewhere between 4 and 5 feet. Since the area of the rug ends in a 5, one of the dimensions must also end with a 5. Let's try 4.5 for our width:

$$4.5 \cdot 9 = 40.5$$

The width of the rug is 4.5 feet, choice C.

ILLUSTRATIVE PROBLEM

Ms. Runyon makes furniture as a hobby. Last year she made 4-legged tables and 3-legged stools as gifts for her family and friends. When she finished, she had used up 37 legs. How many stools might she have made?

    I. 11   II. 7   III. 3   IV. 5

(A) I and II only   (B) I and III only   (C) II, III, and IV only
   (D) I, II, and III only   (E) I, II, III, and IV

ANALYSIS

If we attempt an algebraic solution, we arrive at one single equation with two variables. However, we can also solve the problem by guessing intelligently and testing that guess. Let's make a table to keep track of our guesses and the test results.

Number of tables	1	2	3	4	5	6	7	8	9	10
Number of legs	4	8	12	16	20	24	28	32	36	40
Number of legs left for stools	33	29	25	21	17	13	9	5	1	—
Number of 3-legged stools	11	—	—	7	—	—	3	—	—	—

Ms. Runyon might have made 3, 7, or 11 stools. This is choice D.

ILLUSTRATIVE PROBLEM

In a game at the county fair, each color chip has a different point value: blue = 2 points, yellow = 3 points, and green = 5 points. Amy has collected chips worth 55 points. She has 12 yellow chips. How many green chips might she have?

$$\text{I. } 1 \quad \text{II. } 2 \quad \text{III. } 3 \quad \text{IV. } 4$$

(A) I and II   (B) I and III   (C) II and III   (D) II and IV
(E) I, II, and III

ANALYSIS

Since Amy has 12 yellow chips, we have accounted for 36 points. We now set up an equation for the remaining number of points: $2b + 5g = 19$. However, since there is no second equation, let's use the guess-and-test strategy. The maximum number of blue chips is 9 (since $9 \cdot 2 = 18$), and the maximum number of green chips is 3 (since $3 \cdot 5 = 15$), or we would have more than 19 points. Let's make some intelligent guesses:

Number of green chips (5 points each)	Number of blue chips (2 points each)	Total points
$1 \cdot 5 = 5$	$7 \cdot 2 = 14$	$5 + 14 = 19$
$2 \cdot 5 = 10$	$4 \cdot 2 = 8$	$10 + 8 = 18$ (19 not possible)
$3 \cdot 5 = 15$	$2 \cdot 2 = 4$	$15 + 4 = 19$

Amy could have 3 green or 1 green. The correct answer is choice B.

ILLUSTRATIVE PROBLEM

Lisa bought a scarf for $5 and then spent half of her remaining money on a pair of jogging shoes. Then she spent $2 on a hot dog for lunch. After lunch, she bought a present for her father with half of her remaining money. She was left with $10. How much had she started with?

(A) $85   (B) $50   (C) $49   (D) $48   (E) $45

ANALYSIS

Your students might decide to solve this problem by working backwards (see Strategy 1), since the final condition is given and the earlier status is requested. They might also try to set up an equation and solve it. However, some students might find intelligent guessing and testing an easier procedure to try.

Let's begin with a guess of $85.

$85 - $5 = $80
½($80) = $40
$40 - $2 = $38
½($38) = $19

This is too large, so we must choose a smaller guess. Let's try $45.

$45 - $5 = $40
½($40) = $20
$20 - $2 = $18
½($18) = $9

This is too small, but it appears to be close to our answer. Let's try $49. Notice that we are trying guesses that will be divisible by 2 after we subtract 5. This is the "intelligent" part of the process. This should be emphasized by mentioning that the sum or difference of two odd numbers is always even.

$49 - $5 = $44
½($44) = $22
$22 - $2 = $20
½($20) = $10

The answer is (C) $49.00.

The working-backwards strategy can be applied as follows:

Ended with $10	$10
Before she spent half of her money on her father's present	$20
Before she spent $2 on a hot dog	$22
Before she spent half of her money on jogging shoes	$44
Before she bought a scarf for $5	$49

This problem provides an excellent opportunity to reinforce the idea that problems can often be solved in several different ways. Your students should practice using a variety of solutions as often as possible, in order to become comfortable with all the problem-solving strategies at their disposal.

STRATEGY 8: ACCOUNTING FOR ALL POSSIBILITIES

Students must constantly be reminded that organizing data in a meaningful manner is very important in problem solving on the *SAT I* exam. Indeed, the very construction of a list often reveals patterns within the data that help resolve the problem situation quickly. There is, however, one particular type of list that should be discussed with the students: the exhaustive list, in which the student lists all the possibilities in an organized, systematic way. Somewhere on this list will be the material they are seeking. The exhaustive list is an ideal way to check all the possibilities.

ILLUSTRATIVE PROBLEM

Rebecca and Shawn both work at the local fast food restaurant. Rebecca works one day and then has two days off. Shawn works two days and then has three successive days off. If they both work together on January 1st, on how many days in January will they both work together?

(A) 2   (B) 3   (C) 4   (D) 5   (E) 6

ANALYSIS

Have your students prepare a set of exhaustive lists, showing the dates on which both Rebecca and Shawn work:

Rebecca	1	4	7	10	13	16	19	22	25	28	31		
Shawn	1	2	6	7	11	12	16	17	21	22	26	27	31

Now we can examine the two lists to find what dates they have in common. Both work on the 1st, 7th, 16th, 22nd, and 31st. They work together on 5 days in January. The correct answer is choice D.

ILLUSTRATIVE PROBLEM

Ms. Shuttleworth has three dogs at home. When asked their ages, she replied that the product of their ages is 36. When pressed further, she added this: "The sum of their ages is the same as the age of my son." Her friend stated that she knows the age of Ms. Shuttleworth's son, but still can't tell the ages of the three dogs. "I forgot to tell you," said Ms. Shuttleworth, "the youngest dog is a collie." How old is the youngest dog?

(A) 6   (B) 3   (C) 2   (D) 1   (E) None of these

ANALYSIS

If your students attempt this problem by setting up a system of equations, they will obtain

$xyz = 36$
$x + y + z = $ son's age.

Notice that we now have more variables than we do equations. As a result, this approach appears to lead nowhere. However, we can still apply the technique of preparing an organized, exhaustive list. Have the students begin by listing all the number triples whose product is 36. Notice that the list must be constructed in a systematic, organized way, so as not to miss any of the possibilities:

1 - 1 - 36	2 - 2 - 9	3 - 3 - 4
1 - 2 - 18	2 - 3 - 6	
1 - 3 - 12		
1 - 4 - 9		
1 - 6 - 6		

These eight entries are the complete list. Have the students try other combinations, mentioning that 4 - 3 - 3 is the same as 3 - 3 - 4, and so forth. Now we have an exhaustive list; one of these triples is our answer. Let's move on to the second fact: the *sum* of their ages.

We'll sum the triples in turn:

1 + 1 + 36 = 38	2 + 2 + 9 = 13	3 + 3 + 4 = 10
1 + 2 + 18 = 21	2 + 3 + 6 = 11	
1 + 3 + 12 = 16		
1 + 4 + 9 = 14		
1 + 6 + 6 = 13		

If Ms. Shuttleworth's friend knows the son's age and still can't tell the ages of the dogs, what caused the problem? After all, if the son's age were 21, the dogs' ages would be 1, 2, and 18. If the son's age were 14, the dogs' ages would be 1, 4, and 9. Aha!—the son's age must be 13, since this is the only age that could cause a problem (since there are two sets whose sum is 13). The *youngest* dog is a collie; therefore, the ages must be 6, 6, and 1, for this is the only triple with both a sum of 13 and a single "youngest." The problem is now solved—the answer is (D) 1.

ILLUSTRATIVE PROBLEM

A local movie theater is showing two short films, one in each of two adjoining theaters. The first film runs 36 minutes; the second runs 45 minutes. They both begin at 8:45 a.m. and run continuously until 5:45 p.m. At what times during the day after 8:45 a.m. will both films start at exactly the same time?

I. 11:09 a.m.   II. 11:45 a.m.   III. 2:09 p.m.   IV. 2:45 p.m.

(A) I only   (B) II only   (C) III only
(D) I and III only   (E) II and IV only

ANALYSIS

Students may decide to examine the 36 minutes and 45 minutes to find the common multiples of both. For 36 we find the factors to be 9 · 4; for 45 we find the factors to be 9 · 5. Thus the least common multiple for both would be 9 · 4 · 5 = 180, or exactly every 3 hours. The films start together at 11:45 a.m. and 2:45 p.m. Thus, the correct answer would be choice E.

Students may find, however, that accounting for all the possibilities (probably using a calculator) with an exhaustive list for each would provide another method for solving the problem in a rather revealing manner.

Starting times

Film #1	Film #2
8:45	8:45
9:21	9:30
9:57	10:15
10:33	11:00
11:09	11:45
11:45	12:30
12:21	1:15
12:57	2:00
1:33	2:45
2:09	3:30
2:45	4:15
3:21	5:00
3:57	
4:33	
5:09	

Numbers common to both lists clearly reveal the common starting times to be 11:45 a.m. and 2:45 p.m. So again, we find that the correct answer is choice E.

ILLUSTRATIVE PROBLEM

If a and b are both integers, how many ordered pairs (a, b) will satisfy the equation $a^2 + b^2 = 10$?

(A) 4   (B) 6   (C) 8   (D) 10   (E) None of these

ANALYSIS

Since we are working with integers for a and b, we must look for solutions in which a^2 and b^2 are whole numbers. Since the sum of these squares is 10, any number with an absolute value greater than 3 would give a square greater than 10. We need examine only those perfect squares less than 10, meaning we must examine only $a = 1, 2,$ or 3. If $a = 1$ ($a^2 = 1$) and $b = 3$ ($b^2 = 9$), we have a set that satisfies the equation. The only absolute values that will satisfy the equation are $a = 1, b = 3$, and their symmetric opposites, $a = 3, b = 1$. But since we are dealing with squares, we may also consider both positive and negative answers, which means there are 8 pairs of answers that will satisfy the equation. We list them to be certain that we have accounted for all the possibilities: (1, 3), (1, -3), (-1, 3), (-1, -3), (3, 1), (3, -1), (-3, 1), and (-3, -1). The correct answer is (C) 8.

ILLUSTRATIVE PROBLEM

Four fair coins are tossed. What is the probability that at least two heads are face up?

(A) 11/16   (B) 9/16   (C) 3/16   (D) 4/7   (E) 3/7

ANALYSIS

In order to solve this problem, students may attempt to resort to one or more of the typical permutation, combination, and probability formulas. However, the problem can be easily resolved by making an exhaustive list showing all the possible cases, that is, the entire sample space.

```
HHHH      HHTT      HTTT      TTTT
HHHT      HTHT      THTT
HHTH      THHT      TTHT
HTHH      THTH      TTTH
THHH      TTHH
          HTTH
```

There are 16 possible cases. Of these, we can easily count the number of cases with two or more heads. The correct answer is (A) 11/16.

ILLUSTRATIVE PROBLEM

In the given figure, what is the number of common tangents taken to two circles at a time?

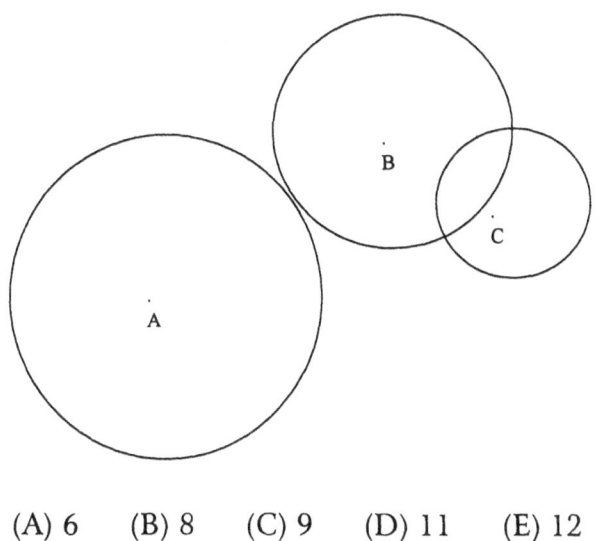

(A) 6   (B) 8   (C) 9   (D) 11   (E) 12

ANALYSIS

While we could draw all of the common tangents and count them, we would not necessarily get them all, since the drawing would probably become too confusing. However, let's take the circles two at a time, and account for all the possibilities.

Circles A and B: 2 external tangents + 1 internal tangent
Circles A and C: 2 external tangents + 2 internal tangents
Circles B and C: 2 external tangents

Thus there are a total of 9 tangents in all, choice C. Emphasize to your students that the problem was easily resolved by accounting for all the possibilities.

STRATEGY 9: ORGANIZING DATA

While it should be axiomatic that students must learn to carefully organize data in a problem, we find that they must constantly be reminded of this. Furthermore, we often find that using a slightly different approach to the method of organization sometimes yields information that leads to a quick resolution of the problem. Thus it becomes imperative that you emphasize to your students the necessity of thinking about how they might organize the data to obtain the most information from it. Particularly in a problem in which there is a great deal of information, organizing the data helps students see what they must do. Notice that this organization can sometimes involve a visual approach.

ILLUSTRATIVE PROBLEM

Given the sequence of integers

$$1, 2, 2, 3, 3, 3, 4, 4, 4, 4, 5, 5, 5, 5, 5, \ldots$$

where each positive integer n occurs in a grouping of n consecutive terms. How many terms are needed so that the sum of the reciprocals is 500?

(A) 1,000,000   (B) 240,650   (C) 125,250   (D) 75,000
(E) None of these

ANALYSIS

At first glance, the problem as posed appears to require some extensive number-crunching. However, let's organize data in a slightly different format. That is, let's follow the wording in the original problem.

Perhaps organizing the data in this way will enable us to solve the problem more readily.

$$\frac{1}{1}, \quad \frac{1}{2} \; \frac{1}{2}, \quad \frac{1}{3} \; \frac{1}{3} \; \frac{1}{3}, \quad \frac{1}{4} \; \frac{1}{4} \; \frac{1}{4} \; \frac{1}{4}, \ldots$$

Now help your students see the pattern that has emerged. Examine the fractions in "clusters":

$$\frac{1}{1} + 1 \; (1 \text{ term})$$

$$\frac{1}{2} + \frac{1}{2} = 1 \; (2 \text{ terms})$$

$$\frac{1}{3} + \frac{1}{3} + \frac{1}{3} = 1 \; (3 \text{ terms})$$

$$\frac{1}{4} + \frac{1}{4} + \frac{1}{4} + \frac{1}{4} = 1 \; (4 \text{ terms}).$$

Thus, for every grouping of n consecutive terms in the original series, the sum of the reciprocals is 1. Now the problem is easily resolved. We are really looking for the sum of the first 500 integers from 1 through 500. The correct answer is (C) 125,250. The problem was easily solved once we had organized the data in a more meaningful manner.

ILLUSTRATIVE PROBLEM

An all-girls' school has 220 students. Of these, 163 play field hockey, 175 play basketball, and 24 play neither sport. How many girls play both sports?

(A) 45   (B) 57   (C) 142   (D) 196   (E) None of these

ANALYSIS

Some students will attempt to solve this problem by simply reasoning it out. However, Venn diagrams are excellent organizational tools to use when classes of objects are being discussed, especially when there is some overlap within those classes. We set up a rectangle to represent the student body (220 girls). One circle represents those who play field hockey (163) and the other circle represents those who play basketball (175). The area of intersection represents those who play both, whereas the portion of the rectangle that is outside of both circles represents those who play neither. Suppose we allow x to represent the number of girls who play both sports.

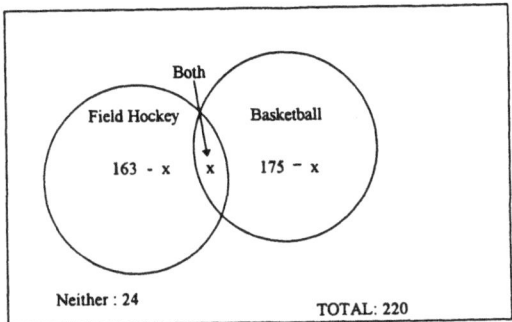

Now we see that 163 - x are those who play only field hockey, while 175 - x represents those who play only basketball. Since the total student population is 220, we obtain the following equations:

$$(163 - x) + (175 - x) + x + 24 = 220$$
$$338 - x = 196$$

Since $x = 142$, choice C is the correct answer.

ILLUSTRATIVE PROBLEM

How many triangles are in the given figure?

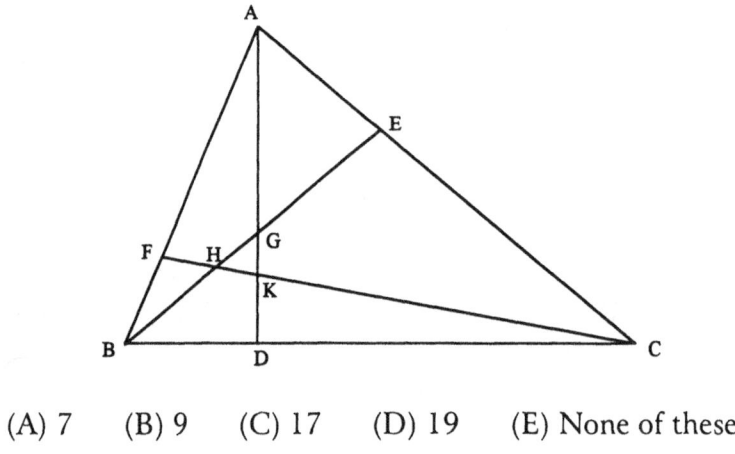

(A) 7 (B) 9 (C) 17 (D) 19 (E) None of these

ANALYSIS

If your students attempt to count the number of triangles in the figure, they will be almost certain to miss some of them in their counting. It is obvious that we need some method of organizing our information to obtain an accurate answer. Let's try to simplify the problem and gradually add the required lines, counting in an organized manner as we go.

Start with the original triangle, ABC. Thus we have exactly 1 triangle.

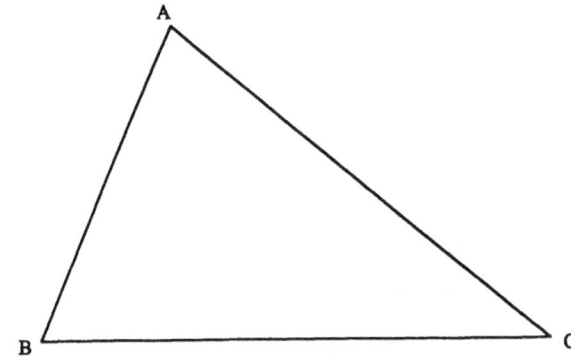

Figure 1.

Now consider △ABC with one interior line, AD. We now have 2 new triangles, ABD and ADC.

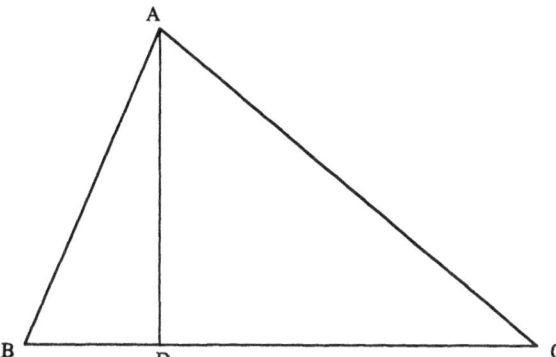

Figure 2.

Now add the next interior line, BE, and count all the new triangles that have all or part of line BE as a side.

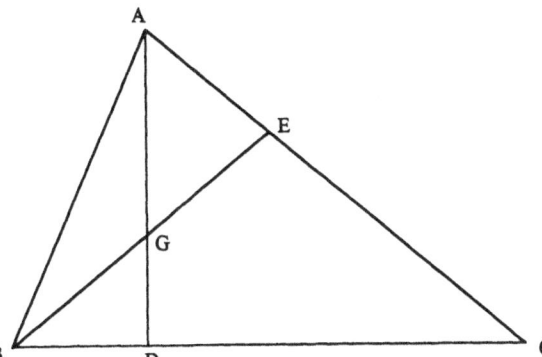

Figure 3.

Continue in this manner, adding line CF. Count the new triangles using all or part of CF as a side.

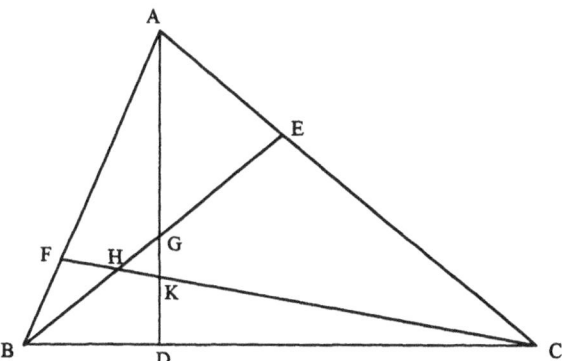

Figure 4.

Let's put these results into a table:

	Added line	New triangles
Figure 1	0	1
Figure 2	AD	2 (ABD, ACD)
Figure 3	BGE	5 (ABG, BGD, AGE, BEC, ABE)
Figure 4	CKHF	9 (FBH, AFC, BHC, AFK, KDC, AKC, FBC, HKG, EHC)
Total		17

There are 17 triangles in the figure, choice C.

Illustrative Problem

Given 8 dots, no 3 of which are on a straight line, how many line segments need to be drawn to connect every pair of dots with a straight line?

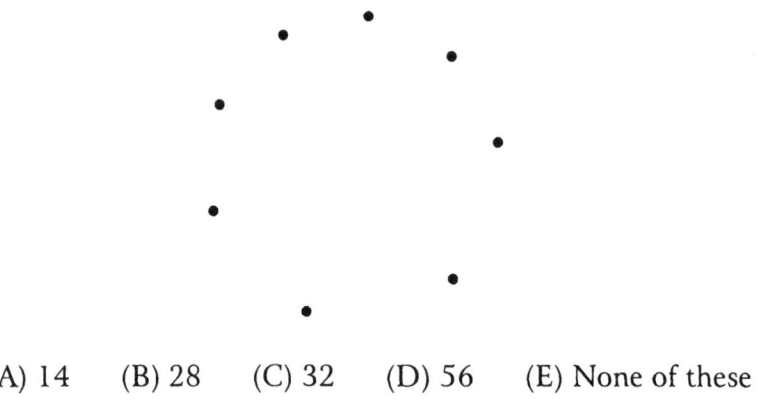

(A) 14   (B) 28   (C) 32   (D) 56   (E) None of these

Analysis

Many of your students may jump right in and begin to draw straight lines connecting the points in pairs. If they try this, they will soon become confused about which points have already been joined. Let's try a different method of organization. First, we'll label the points.

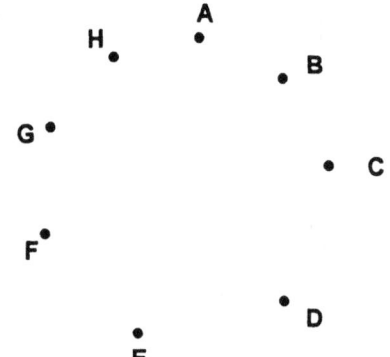

Starting at A, how many line segments are needed to join point A to each of the other points?

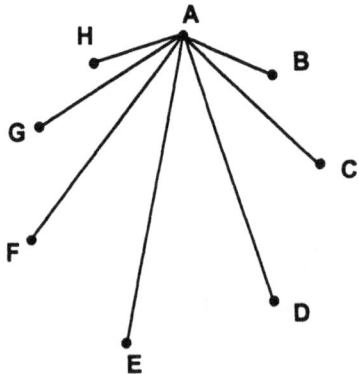

Since there are 7 other points, we need 7 line segments:

$$\overline{AB}, \overline{AC}, \overline{AD}, \overline{AE}, \overline{AF}, \overline{AG}, \overline{AH}.$$

Now consider point B. To join B to every one of the other points, we also need 7 line segments. However, the line segment from B to A is the same line segment as the one already drawn from A to B. Thus we need only 6 line segments to connect B to every other point: $\overline{BC}, \overline{BD}, \overline{BE}, \overline{BF}, \overline{BG}, \overline{BH}$.

If we follow this line of reasoning, it will take

> 5 line segments to join point C to each of the other points
> 4 line segments to join point D to each of the other points
> 3 line segments to join point E to each of the other points
> 2 line segments to join point F to each of the other points
> 1 line segment to join point G to each of the other points.

Point H is already joined to each of the other points. Thus, we need $7 + 6 + 5 + 4 + 3 + 2 + 1 = 28$ line segments to join all the pairs of points. The answer is choice B.

Emphasize to your students that a careful plan of organization often makes solving a complex problem easier than merely jumping in without an organized plan.

Notice that this problem is closely related to the classic "handshake" problem. Students might also wish to solve the problem by considering a smaller number of cases (say 2 points), expanding to 3, 4, and 5, . . . points, and observing a pattern that results; this is another perfectly valid method for solving the problem.

ILLUSTRATIVE PROBLEM

Michael and Eric are about to have a snack consisting of cupcakes. Michael has 3 cupcakes and Eric has 5 cupcakes. Just as they are about to eat, Lauren comes along. Lauren has no cupcakes, but she does have 40 cents to pay for her share. The cupcakes are divided equally among Michael, Eric, and Lauren. When Lauren leaves, Michael and Eric divide up the 40 cents. How much does Eric receive?

(A) 5 cents   (B) 25 cents   (C) 35 cents   (D) 40 cents
(E) None of these

ANALYSIS

At first glance, students might decide to divide the 40 cents in the ratio of 3:5, the number of cupcakes each person contributed. However, this neglects the fact that Michael and Eric each ate 1/3 of the cupcakes. Let's organize the data carefully. Eight cupcakes are placed into the common "pot" to start.

	Michael	Eric	Lauren
Amount put into the pot	3	5	40 cents
Amount eaten from pot	8/3	8/3	8/3
Amount left in pot	1/3	7/3	40 cents

Thus Michael gets 1/8 of the money and Eric gets 7/8, for a 7:1 ratio. Eric receives 35 cents, which is choice C.

STRATEGY 10: DEDUCTIVE REASONING

Deductive reasoning is a strategy your students must master if they are to become good problem solvers. It is a strategy that plays a role when applying most of the other strategies. It is, however, a strategy in its own right, when "reasoning" dominates the problem-solving process. This section will provide examples of problems the solution of which depends largely on reasoning deductively.

ILLUSTRATIVE PROBLEM

In a drawer, there are 8 blue socks, 6 green socks, and 12 black socks. What is the minimum number of socks Henry must take from the drawer, without looking, to be certain that he has 2 socks of the same color?

(A) 2   (B) 4   (C) 6   (D) 8   (E) 12

ANALYSIS

The phrase "*certain* [to have] two socks of the same color" is the key to the problem. The problem does not specify which color, so any of the three would be correct. To solve this problem, have your students reason from a "worst-case scenario." Henry picks one blue sock, one green sock, then one black sock. He now has one of each color, but no matching pair. (True, he might have picked a pair on his first two selections, but the problem calls for certainty.) Notice that as soon as he now picks the fourth sock, he must have a pair of the same color. The correct answer is choice B.

ILLUSTRATIVE PROBLEM

In a drawer there are 8 blue socks, 6 green socks, and 12 black socks. What is the minimum number of socks that Evelyn must take from the drawer, without looking, to be certain that she has 2 black socks?

(A) 2   (B) 4   (C) 8   (D) 12   (E) 16

ANALYSIS

Although this problem appears to be similar to the previous one, there is one important difference. In this problem, a specific color has been selected—it is a pair of *black* socks that we must guarantee being selected. Again, let's use deductive reasoning and construct the "worst-case scenario." Suppose Evelyn first picks all of the blue socks (8). Next she picks all of the green socks (6). Still not one black sock has been chosen. She now has 14 socks in all, but none of them are black. However, the next two socks she picks must be black, since that is the only color remaining. So in order to be *certain* of picking two black socks, Evelyn must select $8 + 6 + 2 = 16$ socks, choice E.

ILLUSTRATIVE PROBLEM

On a shelf in Gladys's basement, there are 3 boxes. One contains only nickels, one contains only dimes, and one contains a mixture of nickels and dimes. The three labels ("Nickels," "Dimes," and "Mixed") fell off, and each was put back on the wrong box. Without looking, Gladys can select one coin from one of the mislabeled boxes and then correctly label all 3 boxes. From which box should Gladys select the coin?

(A) The box labeled "Nickels"
(B) The box labeled "Dimes"
(C) The box labeled "Mixed"
(D) The box labeled "Nickels" or the box labeled "Dimes"
(E) It cannot be done from the given information

ANALYSIS

Students may reason that the "symmetry" of the problem situation dictates that whatever we can say about the box mislabeled "Nickels" could just as well have been said about the box mislabeled "Dimes." Thus, if Gladys chooses a coin from either of these boxes, the results would be the same. This eliminates choices A and B. Students should therefore concentrate their investigations on what happens if we choose from the box mislabeled "Mixed." Suppose Gladys selects a nickel from the Mixed box. Since this box is mislabeled, it cannot be the mixed box and must be, in reality, the Nickel box. Since the box marked "Dimes" cannot really be dimes, it must be the Mixed box. This leaves the third box to be the Dimes box. The correct answer to the problem is choice C.

ILLUSTRATIVE PROBLEM

The picture below shows three views of the same cube, which has a different figure on each face.

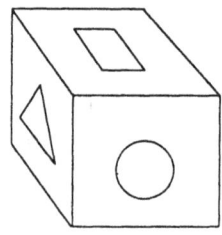

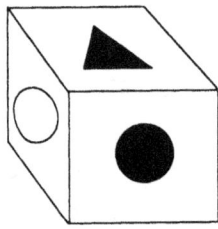

Which figure is opposite the unshaded circle?

(A) The shaded square   (B) The shaded triangle
(C) The unshaded triangle   (D) The unshaded square
(E) The shaded circle

ANALYSIS

Let's make some deductions based on what *cannot* be opposite the unshaded circle. The first view of the cube shows the unshaded square and the unshaded triangle adjacent to the unshaded circle; they cannot be in opposite positions. Similarly, the third view shows the shaded triangle and the shaded circle adjacent to the unshaded circle; they cannot be in opposite positions. This leaves only the shaded square as opposite the unshaded circle. The correct choice is A.

ILLUSTRATIVE PROBLEM

Ms. Stryker is forming committees in her class of 30 children. The committees may be formed with 4, 5, 7, or 8 children. What is the minimum number of committees she can form so that every child is on exactly one committee?

(A) 2   (B) 4   (C) 5   (D) 6   (E) 7

ANALYSIS

Since we are interested in minimizing the number of committees that Ms. Stryker wishes to form, she must try to form as many committees of 8 as possible. Suppose she creates three committees of 8. This accounts for 24 children. The remaining 6 children cannot fit her committee scheme. She now tries using two committees of 8, leaving 14 children to be placed on committees. The least number of committees for these 14 would be two committees of 7 each. The minimum number of committees, therefore, would be 4, choice B.

ILLUSTRATIVE PROBLEM

David and Lisa are playing a game. Whoever loses a match gives the other person a penny. When they were done, David had won 3 times, but Lisa had 8 more pennies than she had started with. How many times did David and Lisa play the game?

(A) 14   (B) 13   (C) 11   (D) 8   (E) None of these

ANALYSIS

Since David won 3 times, Lisa must also have won 3 times just to get back to where she began. Then she won 8 more times. Thus, she must have won a total of 11 matches and lost 3. They played 14 matches—choice A is correct.

ILLUSTRATIVE PROBLEM

The figure below shows three of the faces of a cube. If the six faces of the cube are numbered consecutively, what is the sum of the numbers on all six faces?

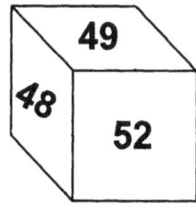

I. 250   II. 297   III. 303

(A) I only   (B) II only   (C) III only
(D) I or II   (E) II or III

Analysis

We know that the six faces of the cube will have six consecutive numbers. We see three of these. Let's make use of some deductive reasoning. Since we see only 48, 49, and 52, there must also be a 50 and 51. These numbers—48, 49, 50, 51, and 52—represent five of the six faces. The sixth number could occur on either end of the sequence. Thus, there are two possibilities for the sixth number, either 47 or 53. This yields two possible sums: 297 or 303. (A calculator comes in handy here.) The correct choice is E.

Illustrative Problem

A farmer plants corn in a square grid so that the number of rows and columns are equal. He increases the size of his corn field equally in the number of rows and columns, to create a new field that contains 211 additional corn plants. How many corn plants were in one row of the original field?

(A) 65   (B) 73   (C) 105   (D) 211   (E) 273

Analysis

If the original field had x corn plants in a row, then the number of corn plants would be x^2. If the new field has h additional plants in each row, then it will contain $(x + h)^2$ plants. Thus,

$$x^2 + 211 = (x + h)^2$$
$$x^2 + 211 = x^2 + 2hx + h^2$$
$$211 = h^2 + 2hx$$
$$211 = h(h + 2x)$$

This appears to be rather complex, since it is a quadratic equation in h, but also contains x. So, let's make use of some intelligent guessing and testing. The number 211 is a prime number. Since h and x must be whole numbers, h must be 1, and $h + 2x$ must be 211. Thus $2x = 210$ and $x = 105$, choice C.

Illustrative Problem

Steve and Al jog around a sports track during their lunch hour. They start at 12:00 noon and run in opposite directions. Steve completes a lap in 5 minutes, whereas Al completes a lap in 4 minutes. Both keep on jogging at the same pace until they meet face to face at the finish line at the exact same time. Who will have run more laps when they meet, and how many more?

(A) Al, 6 laps more   (B) Steve, 5 laps more
(C) Al, 1 lap more   (D) Steve, 2 laps more
(E) You cannot tell from the given information

Analysis

We can make an organized list (see Strategy 8: Accounting for All Possibilities) of all the laps they run:

Al	Steve
12:00	12:00
12:04	12:05
12:08	12:10
12:12	12:15
12:16	12:20
12:20	

They will meet at 12:20. Al will have run 5 laps, whereas Steve will have run 4 laps. The correct answer is choice C. However, this problem and others like it can also be solved with deductive reasoning. The more sophisticated thinker would see that choices B and D cannot hold, because Al is the faster runner. Then, regardless of the direction, we seek the least common multiple of 5 and 4, which is 20. Hence, they will meet in 20 minutes, with Al having run 5 laps and Steve 4 laps.

APPENDIX: REVIEW OF MATHEMATICAL PRINCIPLES

Many problems on the *SAT I* can be solved using simple mathematical operations. However, as we have stated several times, speed and accuracy are crucial factors in students' success on these examinations. It does little good to do a problem if the answer is wrong because of an error in reducing a fraction to lower terms or converting a fraction to decimal form. While it is true that the use of a calculator can ensure accuracy, the correct buttons must be selected, and it often takes more time to use a calculator than it would to do the work mentally. This is just one of the many reasons why it is important that your students spend some time brushing up on their fundamental mathematics skills.

Keep in mind that this review section is merely a refresher of those concepts that should have been acquired by your students in their previous mathematics study. Most of the topics are probably embedded in your students' minds; but, like muscles that begin to atrophy when not in use for a while, these topics may need to be exercised in the brain before they can again be comfortably and proficiently used. In addition, there may be some topics that have not been included in all mathematics curricula. Perhaps the way the topics and concepts are presented here in summary form will not only rekindle them in the students' memory but also firm up a new level of comprehension of the concepts. For example, the summary form of the topic of angle measurement in a circle or the review of the Pythagorean theorem may provide a more complete understanding of these topics. It is also possible that, by the time your students have reached the point where they are preparing for the *SAT I* examination, they may have forgotten some of the very basic concepts from arithmetic. As you develop this course to prepare your students for the *SAT I*, do not omit these basic concepts and skills. They are extremely important and should be an integral part of your total review for the examination.

RULES FOR DIVISIBILITY

It is often very useful to be able to determine what numbers will divide a given number exactly. For example, to reduce the fraction $\frac{2937}{4521}$ to lowest terms, we have to find common divisors for numerator and denominator. To do this by trial and error is a very time-consuming procedure. By using the following rules for divisibility, we may reduce the amount of work considerably.

Rule for divisibility by 2
The last digit of the given number must be even (or 0).

Example
1838 is divisible by 2, since the last digit, 8, is even.

Rule for divisibility by 3
The sum of the digits of the given number must be divisible by 3.

Example
2937 is divisible by 3, since the sum of the digits, $2 + 9 + 3 + 7 = 21$, is divisible by 3.

Rule for divisibility by 4
The last two digits of the given number must form a number that is divisible by 4.

Example
7312 is divisible by 4, since the last two digits form the number 12, which is divisible by 4.

Rule for divisibility by 5
The last digit of the given number must be either 0 or 5.

Example
3785 is divisible by 5, since the last digit is 5.

Rule for divisibility by 6
The rules for divisibility by both 2 *and* 3 must hold, since their product is 6.

Example
1758 is divisible by 2, since the last digit, 8, is even.
    AND
1758 is divisible by 3, since the sum of the digits, $1 + 7 + 5 + 8 = 21$, is a number divisible by 3.
Therefore, 1758 is divisible by 6.

Rule for divisibility by 7
Cut off the last digit from the given number; then subtract twice this deleted digit from the remaining number. The result obtained must be divisible by 7. For large numbers, this procedure may have to be repeated.

Example 1
959 is divisible by 7:

	95 | 9
(Subtract 2 x 9)	<u>18</u>
	77 (which is divisible by 7)

Example 2
5803 is divisible by 7:

	580 | 3
(Subtract 2 x 3)	<u> 6</u>
	574
(Repeat process)	57 | 4
(Subtract 2 x 4)	<u> 8</u>
	49 (which is divisible by 7)

Rule for divisibility by 8
The last three digits of the given number must form a number that is divisible by 8.

Example
57256 is divisible by 8, since the number formed by the last three digits, 256, is divisible by 8.

Rule for divisibility by 9
The sum of the digits of the given number must be divisible by 9.

Example
5382 is divisible by 9, since the sum of the digits, $5 + 3 + 8 + 2 = 18$, is divisible by 9.

Rule for divisibility by 10
The last digit of the given number must be 0.

Example
1970 is divisible by 10, since the last digit is 0.

Rule for divisibility by 11
The difference of the two sums of alternate digits of the given number must be divisible by 11.

Example 1
91927 is divisible by 11, since the difference of the sums of alternate digits, (9 + 9 + 7) - (1 + 2) = 22, is divisible by 11.

Example 2
4521 is divisible by 11, since the difference of the sums of the alternate digits, (4 + 2)-(5 + 1) = 0, is divisible by 11. (Note: Zero is divisible by any nonzero number.)

Rule for divisibility by 12
The rules for divisibility by both 3 *and* 4 must hold, since their product is 12.

Example
1524 is divisible by 3, since 1 + 5 + 2 + 4 = 12, which is a number divisible by 3.
   AND
1524 is divisible by 4, since the last two digits form a number, 24, which is divisible by 4. Therefore, 1524 is divisible by 12.

Rules for divisibility by composites (i.e., nonprimes) greater than 12 follow the same scheme as that for 12 and 6. For example, in order for a given number to be divisible by 15, the rules for divisibility by 3 *and* 5 must both hold.

ORDER OF OPERATIONS

We frequently are faced with the problem of evaluating an expression like the following:
$$5 - 7(5 - 2) + 8 \div 4 + 6$$
If there were no generally accepted rules for the order in which to perform the indicated operations, there would be several possible results. Mathematicians have therefore adopted the following rules for the order of operations:

1. Perform the operations indicated inside each pair of parentheses. If there are parentheses within parentheses, start with the interior ones and work outwards.

2. When all parentheses have been cleared away, perform the multiplications and divisions in order from left to right.

3. Last, perform the additions and subtractions in order from left to right.

Example
$5 - 7(5 - 2) + 8 \div 4 + 6 = ?$

By rule 1, we obtain:
$5 - 7 \times 3 + 8 \div 4 + 6$

By rule 2, we obtain:
$5 - 21 + 2 + 6$

By rule 3, we obtain:
-8

A Brief Review of Fractions, Decimal Fractions, and Percents

1. Parts of a fraction
In the fraction $\frac{3}{5}$, the *numerator* is 3, and the *denominator* is 5.

2. Proper and improper fractions
If the numerator of a fraction is less than the denominator, then it is a proper fraction (e.g., $\frac{3}{4}$). If the denominator of a fraction is less than the numerator, then it is an improper fraction (e.g., $\frac{5}{3}$).

3. Mixed numbers
A mixed number is one in which a whole number and a proper fraction are combined (e.g., $5\frac{2}{3}$). In order to facilitate computation, it is often advisable to change a mixed number to an improper fraction.

Example

To change $5\frac{2}{3}$ to an improper fraction, multiply 5 by 3 and add 2 to the product to get the new numerator, and use the original denominator (3). In this way, $5\frac{2}{3} = \frac{17}{3}$.

4. Reducing fractions to lowest terms
To reduce a fraction to lowest terms, divide both numerator and denominator by a common factor (where possible). This procedure is to be repeated until no common factor exists. To facilitate the search for a common factor, see the section on *"Rules for Divisibility,"* pages 85 - 87.

Example

To reduce $\frac{2937}{4521}$ to lowest terms, divide both numerator and denominator by 3 and 11 (or by their product, 33) to get $\frac{89}{137}$.

5. Addition of fractions
In order to add fractions, we must change each of the fractions to equivalent fractions with the same denominator (preferably the lowest common denominator). Add the numerators (to get the numerator of the sum).

Example

$$\frac{2}{3} + \frac{5}{8} + \frac{7}{12} = ?$$

Since the lowest common denominator of the three denominators (3, 8, and 12) is 24, we can change each fraction to an equivalent fraction with a denominator of 24:

$$\frac{2}{3} = \frac{2}{3} \cdot \frac{8}{8} = \frac{16}{24}$$
$$\frac{5}{8} = \frac{5}{8} \cdot \frac{3}{3} = \frac{15}{24}$$
$$\frac{7}{12} = \frac{7}{12} \cdot \frac{2}{2} = \frac{14}{24}$$

By addition, we get: $\frac{16 + 15 + 14}{24} = \frac{45}{24} = \frac{15}{8} = 1\frac{7}{8}$

6. Subtraction of fractions

In order to subtract one fraction from another, we must change each of the fractions to equivalent fractions with a common denominator (as in addition). We must then subtract the numerators (keeping the common denominator).

Example

Subtract $\frac{2}{3}$ from $\frac{4}{5}$.

Since the lowest common denominator of the two denominators (3 and 5) is 15, we must change each fraction to an equivalent fraction with denominator of 15:

$$\frac{4}{5} = \frac{4}{5} \cdot \frac{3}{3} = \frac{12}{15}$$

$$\frac{2}{3} = \frac{2}{3} \cdot \frac{5}{5} = \frac{10}{15}$$

Subtracting the numerators, we get:

$$\frac{12-10}{15} = \frac{2}{15}$$

7. Multiplication of fractions

To multiply fractions, multiply the numerators to get the numerator of the product, and multiply the denominators to get the denominator of the product. To simplify the operation, divide out any common factors between numerators and denominators before doing the multiplications

Example

Multiply $\frac{15}{22}$ by $\frac{33}{65}$.

Divide numerator and denominator by 5, the common factor of 15 and 65:

$$\frac{\overset{3}{\cancel{15}}}{22} \cdot \frac{33}{\underset{13}{\cancel{65}}} = \frac{3}{22} \cdot \frac{33}{13}$$

Divide numerator and denominator by 11, the common factor of 22 and 33:

$$\frac{3}{\underset{2}{\cancel{22}}} \cdot \frac{\overset{3}{\cancel{33}}}{13} = \frac{3}{2} \cdot \frac{3}{13} = \frac{9}{26}$$

8. Division of fractions

To divide one fraction by another, invert the divisor and then multiply the dividend by the inverted divisor.

Example

Divide $\frac{2}{3}$ by $\frac{6}{7}$.

$$\frac{2}{3} \div \frac{6}{7} = \frac{\overset{1}{\cancel{2}}}{3} \cdot \frac{7}{\underset{3}{\cancel{6}}} = \frac{7}{9}$$

9. Comparing fractions

$a/b > c/d$ if and only if $ad > bc$, or $a/b < c/d$ if and only if $ad < bc$.

10. Addition of decimal fractions
The decimal fractions to be added must be placed in a column so that all the decimal points are "lined up."

Example

.007 + 7.3 + 56 + 10.502 = ?
Write the numbers to be added in a column:
```
  .007
 7.3
56.
10.502
73.809
```

11. Subtraction of decimal fractions
The decimal fractions to be subtracted must be placed with the decimal points "lined up."

Example

Subtract 5.2 from 2.05.
```
  5.20
- 2.05
  3.15
```

12. Multiplication of decimal fractions
Multiply the numbers without regard to the decimal points. Then place the decimal point in the product so that there are as many digits to the right of it as the total number of digits to the right of the decimal point in the original numbers. A final zero in the product counts as a digit.

Example

Multiply 5.805 by 3.02.
```
   5.805   (3 digits to the right of the decimal point)
   3.02    (2 digits to the right of the decimal point)
  11610
  17415
  17.53110 (5 digits to the right of the decimal point)
```

13. Division of decimal fractions
(1) Move decimal point to right of final digit of divisor. (2) Move decimal point as many places to right in dividend as it was moved in divisor. (Add zeros, if necessary, to dividend.) (3) Divide by long division. (4) Place decimal point in quotient directly above displaced decimal point in dividend.

Example

Divide 12.243 by 2.31.
```
              (4)
               ∨
              5.3
    2.31.  12.24.3
              11 55
                69 3
                69 3
```

14. Converting a fraction to decimal form
Divide the numerator by the denominator.

Example

To convert $\frac{1}{8}$ to decimal form, divide 1.000 by 8 to get .125.

15. Repeating decimals
Fractions convert either to terminating decimals, as in Rule 14 (above), or to repeating decimals.

Examples

$$\frac{1}{3} = 0.33\ldots, \text{ often written as } 0.\overline{3}$$

$$\frac{1}{11} = 0.0909\ldots \text{ or } 0.\overline{09}$$

$$\frac{611}{4950} = 0.12343434\ldots \text{ or } 0.12\overline{34}$$

16. Converting a repeating decimal to a fraction
Determine the fraction represented by the repeating decimal by following the procedure in the examples below:

$0.333\ldots$ has *one* repeating digit; let $x = 0.33\ldots$;
multiply x by 10 *once* to get $10x = 3.333\ldots$; set up the subtraction:

$$\begin{aligned} 10x &= 3.33\ldots \\ -x &= 0.33\ldots \\ \hline 9x &= 3 \end{aligned}$$

$$x = \frac{3}{9} = \frac{1}{3}$$

$0.0909\ldots$ has *two* repeating digits; let $x = 0.0909\ldots$;
multiply x by 10 *twice* to get $100x = 9.0909\ldots$; set up the subtraction:

$$\begin{aligned} 100x &= 9.0909\ldots \\ -x &= 0.0909\ldots \\ \hline 99x &= 9 \end{aligned}$$

$$x = \frac{9}{99} = \frac{1}{11}$$

$0.123434\ldots$ has *two* repeating decimals; let $x = 0.123434\ldots$;
multiply x by 10 *twice* to get $100x = 12.3434\ldots$; set up the subtraction:

$$\begin{aligned} 100x &= 12.343434\ldots \\ -x &= 0.123434\ldots \\ \hline 99x &= 12.22 \end{aligned}$$

$$x = \frac{12.22}{99} = \frac{1222}{9900} = \frac{611}{4950}$$

17. Converting a decimal fraction to percent form

Multiply by 100%; that is, move the decimal point two places to the right and add the % sign.

Example

To change .35 to percent form, move the decimal point two places to the right (.35.) to get 35%.

18. Converting a fraction to percent form

Multiply by 100%, or convert the fraction to decimal form and then to percent form. (See Rules 14 and 17.)

Example

Change $\frac{3}{8}$ to percent form.

$$\frac{3}{8} \cdot 100\% = \frac{300}{8}\% = 37.5\% \text{ or } \frac{3}{8} = .375 = 37.5\%$$

PERCENTAGE PROBLEMS

In order to simplify the following discussion, let us define the following terms:

Percentage is the result obtained when a percent is taken of a number.
Base is the number of which a percent is being taken.
Rate is the percent taken of a number (i.e., the base).

Following are the three basic types of percentage problems:

Type of Problem	Sample Problems	
Finding percentage: Percentage = Rate × Base	1. Find 35% of 20. **Solution:** (.35)(20) = 7	2. Find 42% of 150. **Solution:** (.42)(150) = 63
Finding the base: Base = $\frac{\text{Percentage}}{\text{Rate}}$	1. 35% of what number is 7? **Solution:** $\frac{7}{.35} = \frac{700}{35} = 20$	2. 42% of what number is 63? **Solution:** $\frac{63}{.42} = \frac{6300}{42} = 150$
Finding the rate: Rate = $\frac{\text{Percentage}}{\text{Base}}$	1. 7 is what percent of 20? **Solution:** $\frac{7}{20} = .35 = 35\%$	2. 63 is what percent of 150? **Solution:** $\frac{63}{150} = .42 = 42\%$

EQUIVALENT FRACTIONS AND PERCENTS YOU SHOULD KNOW

$\frac{1}{2} = 50\%$	$\frac{1}{3} = 33\frac{1}{3}\%$ $\frac{2}{3} = 66\frac{2}{3}\%$	$\frac{1}{4} = 25\%$ $\frac{3}{4} = 75\%$	$\frac{1}{5} = 20\%$ $\frac{2}{5} = 40\%$ $\frac{3}{5} = 60\%$ $\frac{4}{5} = 80\%$	$\frac{1}{6} = 16\frac{2}{3}\%$ $\frac{5}{6} = 83\frac{1}{3}\%$	$\frac{1}{8} = 12\frac{1}{2}\%$ $\frac{3}{8} = 37\frac{1}{2}\%$ $\frac{5}{8} = 62\frac{1}{2}\%$ $\frac{7}{8} = 87\frac{1}{2}\%$
$\frac{1}{7} = 14\frac{2}{7}\%$		$\frac{1}{9} = 11\frac{1}{9}\%$		$\frac{1}{11} = 9\frac{1}{11}\%$	$\frac{1}{12} = 8\frac{1}{3}\%$

A Brief Review of Some Essentials of Algebra

1. Addition of signed numbers

It is important to understand the difference in meaning between a plus or minus sign used to indicate the operation "add" or "subtract," and the same signs used to indicate whether a number is positive or negative. For example, the expression:

$$5 + (-7)$$

means "add the negative number -7 to the positive number 5." (Note that the plus sign is usually omitted before positive numbers.) The expression:

$$5 - 7$$

means "subtract the (positive) number 7 from the (positive) number 5." (Note again the omission of the plus signs.)

A. To add numbers that all have the same sign, add the numbers (absolute values) and keep the same sign.

Example
$(-5) + (-22) + (-7) = -(5 + 22 + 7) = -34$

B. To add numbers with different signs, find the sum of the positive numbers and the sum of the negative numbers. Then find the difference between the absolute values of those two sums. Give the result the sign of the larger of the two sums.

Example
Add: -12, -7, +15, -10, +3, +18.
Add separately the positive and negative numbers:
$(+15) + (+3) + (+18) = +(15 + 3 + 18) = +36$
$(-12) + (-7) + (-10) = -(12 + 7 + 10) = -29$
Find the difference between the absolute values, 36 and 29:
$36 - 29 = 7$
Use the sign of the larger of the two absolute values:
+7

2. Subtraction of signed numbers

To subtract one signed number from another, change the sign of the subtrahend (the number being subtracted) and follow the rules for addition of signed numbers.

Example 1
Subtract -58 from +35.
$+35 - (-58) = +35 + (+58) = +(35 + 58) = +93$

Example 2
Subtract +58 from +35.
$+35 - (+58) = +35 + (-58) = -(58 - 35) = -23$

Example 3
Subtract -58 from -35.
$-35 - (-58) = -35 + (+58) = +(58 - 35) = +23$

3. Multiplication of signed numbers

To multiply two signed numbers, multiply their absolute values; then,
(a) if the signs are alike, give the product a plus sign;
(b) if the signs are unlike, give the product a minus sign.

Examples

$$(+3)(-5) = -15$$
$$(-3)(+5) = -15$$
$$(-3)(-5) = +15$$
$$(+3)(+5) = +15$$

Notes

(a) When multiplying an *odd* number of negative numbers, the product is *negative*.

Example

$$(+5)(-2)(-3)(+2)(-1) = -60$$

(b) When multiplying an *even* number of negative numbers, the product is *positive*.

Example

$$(-5)(-2)(-3)(+2)(-1) = +60$$

It follows that an odd power of a negative number yields a negative result, whereas an even power of a negative number yields a positive result.

4. Division of signed numbers

When dividing a pair of signed numbers, divide their absolute values; then apply the rule of signs for multiplication.

Examples

$$(+15) \div (-3) = -5$$
$$(-15) \div (+3) = -5$$
$$(-15) \div (-3) = +5$$
$$(+15) \div (+3) = +5$$

5. Positive exponents

$$a^n = \underbrace{a \bullet a \bullet a \bullet \cdots \bullet a}_{n \text{ times}}$$

The base is a and the exponent is n. The exponent indicates the number of times the base appears as a factor in the product.

Examples

$$2^1 = 2 \qquad 2^3 = 2 \bullet 2 \bullet 2 = 8 \qquad (-2)^4 = (-2) \bullet (-2) \bullet (-2) \bullet (-2) = 16$$

But $\qquad -(2)^4 = -(2) \bullet (2) \bullet (2) \bullet (2) = -16$

6. Negative exponents

$$a^{-n} = \frac{1}{a^n}$$

A negative exponent indicates that a reciprocal is to be taken, using the same base and the negative of the original exponent.

Examples

$$3^{-1} = \frac{1}{3} \qquad\qquad 3^{-2} = \frac{1}{3^2} = \frac{1}{9}$$

$$(-1)^{-4} = \frac{1}{(-1)^4} = \frac{1}{1} = 1 \qquad\qquad -1^{-4} = -\frac{1}{1^4} = -\frac{1}{1} = -1$$

7. Zero exponent
Any number, except zero, raised to an exponent of zero is equal to 1.

$$a^0 = 1, \quad a \neq 0$$

Zero raised to an exponent of zero is undefined. (See Rule 10.)

Examples

$$3^0 = 1 \qquad\qquad (-4)^0 = 1 \qquad\qquad [\,0^0 \text{ is undefined}\,]$$
$$\text{But } -4^0 = -(4)^0 = -1$$

8. Monomials (coefficients and variables)
The expression

$$ax^n$$

is a monomial. The coefficient a represents a given numerical value; the base x represents an unknown or variable.

Examples

$2x$ 2 is the coefficient

$-x$ -1 is the coefficient

$\dfrac{y}{3}$ $\dfrac{1}{3}$ is the coefficient

9. Multiplying monomials
Multiply the coefficients and add the exponents of factors with the same base.

$$(ax^n) \cdot (bx^m) = abx^{n+m}$$

Examples

$$(2x^3) \cdot (3x^2) = 6x^{3+2} = 6x^5 \qquad\qquad (-5x) \cdot (x) \cdot (9x^2) = -45x^{1+1+2} = -45x^4$$
$$(-5x^2y^3) \cdot (3xy^2) = -15x^3y^5$$

10. Dividing monomials
Divide the coefficients, and for factors with the same base, subtract the exponent of the denominator from the exponent of the numerator.

$$\frac{x^n}{x^m} = x^{n-m}$$

Recall from Rule 7 that $x^0 = 1$. From the rule for dividing monomials:

$$1 = \frac{x^n}{x^n} = x^{n-n} = x^0$$

Examples

$$\frac{8x^3}{2x^2} = 4x^{3-2} = 4x^1 = 4x \qquad\qquad \frac{-2x^4}{10x^7} = \frac{-1x^{4-7}}{5} = \frac{-1x^{-3}}{5} = \frac{-1}{5x^3}$$

11. Raising an exponent to an exponent
Raise the coefficient to the indicated exponent. Multiply the exponents to get the variable's new exponent.

$$(ax^n)^m = a^m x^{nm}$$

Examples

$$(x^2)^3 = x^{2 \cdot 3} = x^6 \qquad\qquad (4x^3)^3 = 4^3 x^{3 \cdot 3} = 64 x^9$$

12. Fractional (rational) exponents
The expression

$$x^{m/n}$$

indicates that the nth root of x is to be taken and the result raised to the mth power:

$$x^{m/n} = \left(x^{1/n}\right)^m = \left(\sqrt[n]{x}\right)^m$$

Examples

$$4^{1/2} = \sqrt{4} = 2 \qquad\qquad 8^{2/3} = \left(8^{1/3}\right)^2 = \left(\sqrt[3]{8}\right)^2 = 2^2 = 4$$

$$25^{-1/2} = \frac{1}{25^{1/2}} = \frac{1}{\sqrt{25}} = \frac{1}{5}$$

13. Like terms
Terms having the same variable and the same exponent are like terms.

Examples

$$3x \text{ and } -4x, \qquad 5n, \frac{n}{2} \text{ and } -n, \qquad -7xy \text{ and } 3xy, \qquad 2 \text{ and } \frac{-1}{6}$$

Notice that constants are like terms whose variable has an exponent of zero:
$$4x^0 = 4 \cdot 1 = 4$$

14. Adding like terms (monomials)
Add (combine) only like terms by adding their coefficients and leaving the bases and exponents as they were.

Example
Combine like terms:
$$-5x^2 + 3xy + 2x + 15x^2 + 7xy - 3x = 10x^2 + 10xy - x$$

15. Subtraction of polynomials
See Rules 2 (page 93) and 14 (above).

Example
Subtract $+12x^2 - 7x$ from $-3x^2 - 9x$.

$$\begin{array}{l} -3x^2 - 9x \\ \underline{-12x^2 + 7x} \\ -15x^2 - 2x \end{array}$$ (Change each sign of the subtrahend and follow the rules for the addition of signed numbers, page 93.)

16. Multiplication of a monomial by a polynomial (removing parentheses)
Multiply the monomial by each term of the polynomial (distributive property), and apply Rule 9 (page 95) to each
of the multiplications.

Example
$-3x^2y^3(5x^2y - 2xy^2 - 1)$
$= -3x^2y^3(5x^2y) - 3x^2y^3(-2xy^2) - 3x^2y^3(-1)$
$= -15x^4y^4 + 6x^3y^5 + 3x^2y^3$

17. Multiplication of a polynomial by a polynomial
Multiply each term of one polynomial by each term in the other polynomial and then collect like terms.

Example
$(3x - 2)(5x^2 + 7x - 1)$
$= 3x(5x^2 + 7x - 1) - 2(5x^2 + 7x - 1)$
$= 15x^3 + 21x^2 - 3x - 10x^2 - 14x + 2$
$= 15x^3 + 11x^2 - 17x + 2$

18. Multiplication of two binomials

In the multiplication of two binomials, the procedure of Rule 17 can be summarized in the following four steps: (1) Multiply the pair of first members; (2) multiply the outer pair; (3) multiply the inner pair; (4) multiply the pair of second members. Then combine like terms (if any).

Example

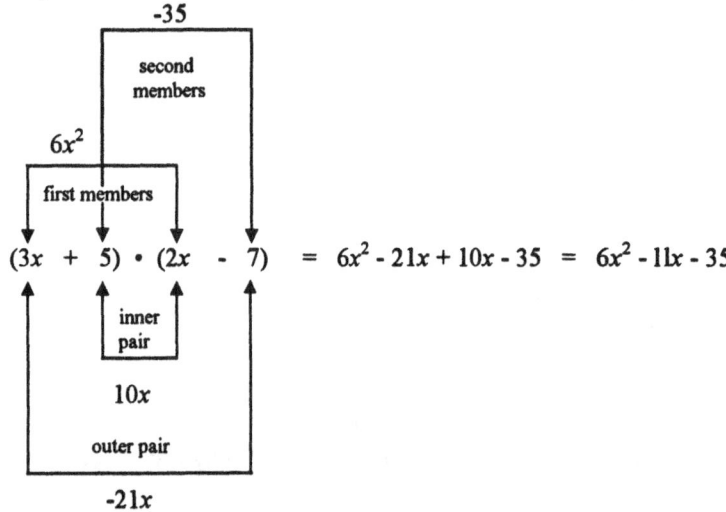

$(3x + 5) \cdot (2x - 7) = 6x^2 - 21x + 10x - 35 = 6x^2 - 11x - 35$

Note: To remember which terms to multiply, remember the word FOIL (First, Outer, Inner, Last).

19. Factoring

There are three frequent factoring situations in algebraic expressions:

A. Common factor in a polynomial:
$18xy^3z - 6x^2y + 9xyz^2 = 3xy(6y^2z - 2x + 3z^2)$

B. Difference of two perfect squares $[a^2 - b^2 = (a + b)(a - b)]$:
$16x^6 - 9y^2 = (4x^3 - 3y)(4x^3 + 3y)$

C. Trinomial:
$6x^2 - x - 15 = (2x + 3)(3x - 5)$
[Factoring a trinomial of this type is a trial-and-error process in which you try various combinations of the factors of the first and last terms, testing each combination by Rule 10 (page 96). There may not be any rational factors.]

20. Evaluating a formula

Merely substitute the appropriate values for each of the variables.

Example

The formula for the volume of a cone is:

$$V = \frac{1}{3} \pi r^2 h$$

Find V, if $\pi = \frac{22}{7}$, $r = 7$, and $h = 12$.

$$V = \frac{1}{3} \cdot \frac{22}{7} \cdot (7)^2 \cdot (12)$$

$$V = (22) \cdot (7) \cdot (4) = 616$$

21. Solving first-degree equations

Example 1
Solve for x:

$51 - 5x = 5x + 9 - 4x$

Combine like terms on each
side of the equation: $51 - 5x = x + 9$

Add $+5x$ to both sides: $\underline{+5x = +5x}$

$51 = 6x + 9$

Add -9 to both sides: $\underline{-9 = -9}$

$42 = 6x$

Multiply both sides by $\frac{1}{6}$: $\frac{1}{6}(42) = \frac{1}{6}(6x)$

$7 = x$

Example 2
Solve for x:

$$\frac{x}{3} + 7 = \frac{x}{5} - 3$$

Multiply both sides of the equation by the lowest common denominator (15):

$15\left(\frac{x}{3}\right) + 15(7) = 15\left(\frac{x}{5}\right) - 15(3)$

or:

$5x + 105 = 3x - 45$

Add $(-3x)$ to both sides: $\underline{-3x = -3x}$

$2x + 105 = -45$

Add (-105) to each side: $\underline{ -105 = -105}$

$2x = -150$

Multiply both sides by $\frac{1}{2}$: $\frac{1}{2}(2x) = \frac{1}{2}(-150)$

$x = -75$

Example 3
Solve for x:

$$\frac{15}{4x} = \frac{1}{8}$$

When there is only one fraction on either side of the equation, we may cross multiply:

$$\frac{15}{4x} \diagup\!\!\!\!\diagdown \frac{1}{8}$$

$(4x)(1) = (15)(8)$

$x = 30$

Note: In this case, one may divide (cancel) by a common factor in any of the following directions:

$$\uparrow\downarrow \frac{a}{b} \leftrightarrow \overset{=}{\leftrightarrow} \frac{c}{d} \uparrow\downarrow$$

$$\frac{15}{4x} = \frac{1}{8}$$
$x 2$
$x = 30$

22. Solving a pair of linear equations simultaneously

Example
Solve for x and y:

$5x - 2y = 8$ (1)
$3x - y = 5$ (2)

Method 1: Substitution

Solve for y in Equation (2):
$y = 3x - 5$ (3)

Substitute this value for y in Equation (1):
$5x - 2(3x - 5) = 8$

Solve for x:
$5x - 6x + 10 = 8$
$-x + 10 = 8$
$-x = -2$
$x = 2$

Substitute this value of x in Equation (3) and solve for y:
$y = 3(2) - 5$
$y = 1$

Check: Substitute $x = 2$ and $y = 1$

in Equation (1)	in Equation (2)
?	?
$5(2) - 2(1) = 8$	$3(2) - 1 = 5$
$8 = 8$	$5 = 5$

Method 2: Addition

$5x - 2y = 8$ (1)
$3x - y = 5$ (2)

Multiply one (or both) equations by factors that will result in a pair of like terms with the same coefficient (absolute value) and opposite signs. An easy way to do this in this case is to multiply Equation (2) by -2:

$-2(3x - y) = -2(5)$
$-6x + 2y = -10$ (3)

Add Equations (1) and (3):
$5x - 2y = 8$
$\underline{-6x + 2y = -10}$
$-x = -2$
$x = 2$

Substitute this value of x in Equation (2):

$3(2) - y = 5$
$6 - y = 5$
$y = 1$

23. Solving a second-degree equation (quadratic)

In order to solve a (factorable) quadratic equation, collect all terms on one side of the equation and have zero on the other side, factor the nonzero side, and set each factor equal to zero. Then solve each linear equation to get two solutions for the variable. (Note: In order for a product $ab = 0$, either $a = 0$, or $b = 0$.)

Example
Solve for x:
$6x^2 + 9x = 6$

Set the terms equal to zero (i.e., add -6
to both sides): $\qquad\qquad\qquad\qquad\qquad 6x^2 + 9x - 6 = 0$
Factor the trinomial: $\qquad\qquad\qquad\qquad\qquad\quad (2x - 1)(3x + 6) = 0$
Set each factor equal to zero: $\qquad\qquad\qquad\qquad 2x - 1 = 0 \;\|\; 3x + 6 = 0$
Solve each linear (first degree) equation: $\qquad\qquad 2x = 1 \;\;\;\;\|\; 3x = -6$
$\qquad\qquad\qquad\qquad\qquad\qquad\qquad\qquad\qquad\quad x = \dfrac{1}{2} \;\;\|\; x = -2$

24. Properties of inequalities

For real numbers a, b, m, n, and k, the following properties are true:

A. If $m > n$ and $a \geq b$, then $m + a > n + b$.

B. If $m > n$, then (1) $m - a > n - a$, and (2) $a - m < a - n$.

C. If $m > n$ and $a > 0$, then $am > an$.

D. If $m > n$ and $a > 0$, then (1) $\dfrac{m}{a} > \dfrac{n}{a}$, and (2) $\dfrac{a}{m} < \dfrac{a}{n}$.

E. If $n = m + k$ and $k > 0$, then $n > m$.

F. If a, b, m, and n are positive, and $\dfrac{a}{b} > \dfrac{m}{n}$, then $an > bm$.

25. Solving linear inequalities

Basically, a linear inequality is solved like a linear equation, with one major *exception*: When both sides of an inequality are multiplied or divided by a negative number, the order of the inequality changes.

Example
Solve for x:
$5 - 3x > 17$
Add -5 to both sides of the inequality to get:
$-3x > 12$
Multiply both sides of the inequality by $-\dfrac{1}{3}$ to get:
$x < -4$

Note: In the last step (which involved multiplication by a negative number) the *order* of the inequality changed.

26. x,y-coordinate system

Number lines intersect at right angles to form the x- and y-axes and the four quadrants of the x, y-coordinate system (also called the Cartesian plane). The quadrants are numbered as shown.

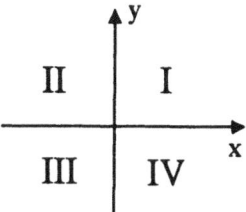

27. Ordered pair
A point on the x,y-plane is named as an ordered pair, with its *x*-coordinate first and its *y*-coordinate second.
$$(x, y)$$

The *x*- and *y*-axes intersect at the point (0, 0), the origin.

Examples

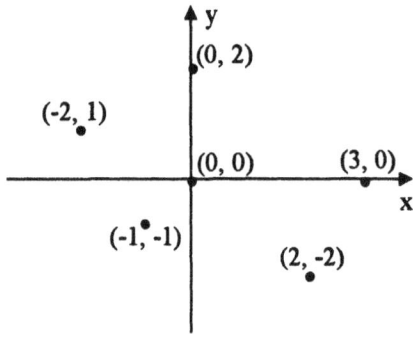

28. Linear equations
Linear equations contain the variables *x* and/or *y*; in a linear equation, the exponent of *x* and *y* is 1, and the *x* and *y* terms are added or subtracted. One way to represent a linear equation is:

$$Ax + By = C$$

in which *A*, *B*, and *C* are constants.

Examples

$$3x + y = 1 \qquad -x = 2 \qquad 7y = 3 \qquad y = -x + 4$$

Examples of equations that are *not* linear:

$$xy = 1 \qquad x^2 + y^3 = 4 \qquad \frac{1}{x} = 2y$$

29. Graphing a linear equation by finding two points

As its name implies, a linear equation represents a line. Two points are needed to graph any line. To determine two points (x_1, y_1) and (x_2, y_2) from a given linear equation, proceed as in the following example.

Example

First solve the equation for y in terms of x:
$$4x + y = 2$$
$$y = -4x + 2$$

Then substitute any value in place of x and calculate y. For example, let $x = 0$, so
$$y = -4\cdot(0) + 2 = 2$$

Thus $(x_1, y_1) = (0, 2)$.

Next, choose another value for x and calculate a second value for y.
For example, let $x = -1$, so
$$y = -4\cdot(-1) + 2 = 6$$

Thus $(x_2, y_2) = (-1, 6)$.

Graph the points $(0, 2)$ and $(-1, 6)$ on the x, y-plane and draw the line through these points.

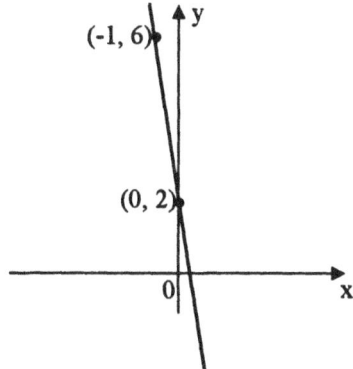

30. Midpoint of a line segment

Given the line segment whose endpoints are (x_1, y_1) and (x_2, y_2), the midpoint of the line segment is given by:

$$(x_m, y_m) = \left(\frac{x_1 + x_2}{2}, \frac{y_1 + y_2}{2}\right)$$

Notice that x_m is the average of x_1 and x_2. Likewise, y_m is the average of y_1 and y_2.

Example

The midpoint of the line segment whose endpoints are $(6, -5)$ and $(-2, 0)$ is

$$(x_m, y_m) = \left(\frac{6 + (-2)}{2}, \frac{-5 + 0}{2}\right) = \left(2, \frac{-5}{2}\right)$$

31. Distance between two points (length of a line segment)

The distance between two points (x_1, y_1) and (x_2, y_2) is given by:

$$d = \sqrt{(x_1 - x_2)^2 + (y_1 - y_2)^2}$$

This formula derives from the fact that a right triangle constructed from the two given points has leg lengths equal to $|x_1 - x_2|$ and $|y_1 - y_2|$, as shown:

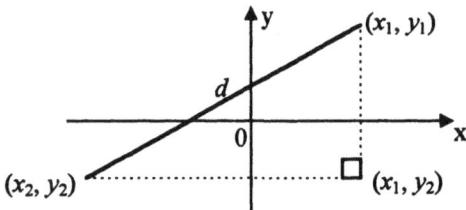

Then, from the Pythagorean theorem,

$$d^2 = (x_1 - x_2)^2 + (y_1 - y_2)^2$$

in which d is the length of the hypotenuse. Solving for d gives the distance formula.

32. Slope of a line

For a line containing the points (x_1, y_1) and (x_2, y_2), the slope of the line is the ratio:

$$m = \frac{y_1 - y_2}{x_1 - x_2}$$

Example

If $(x_1, y_1) = (0, 7)$ and $(x_2, y_2) = (-3, 5)$, then $m = \frac{7-5}{0-(-3)} = \frac{2}{3}$.

33. Positive slope

A line with a positive slope has the general appearance:

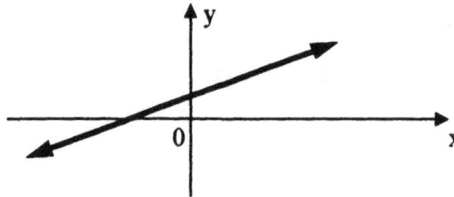

34. Negative slope

A line with a negative slope has the general appearance:

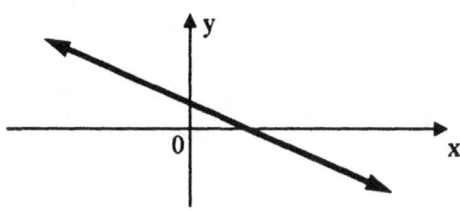

35. Zero slope
A horizontal line has a slope of zero:

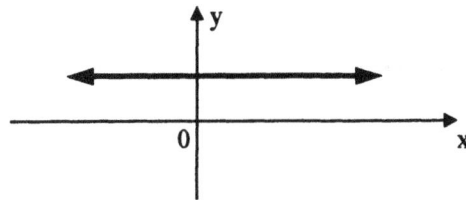

36. No slope (or undefined slope)
A vertical line has no slope:

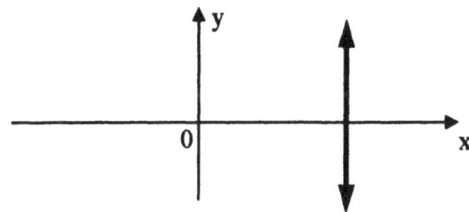

37. Slope-intercept form of a line
Solving a linear equation for y in terms of x gives the slope-intercept form of the line:

$$y = mx + b$$

in which m, the coefficient of x, is the *slope of the line*, and b, the value of y when $x = 0$, is the line's *y-intercept*:

$$y = m \cdot (0) + b = 0 + b = b$$

Examples

Solving $2x + 3y = 12$ for y gives:

$$y = \frac{-2}{3} \cdot x + 4$$

Slope $m = \frac{-2}{3}$ and y-intercept $b = 4$.

Solving $x + y = 6$ for y gives:

$$y = -x + 6$$

in which $m = -1$ and $b = 6$.

38. Graphing the slope-intercept form of a line

Plot the y-intercept $(0, b)$ first. Use the slope m to arrive at a second point.
Starting at $(0, b)$:

If m is positive, move *up* the number of spaces in m's numerator and *right* the number of spaces in m's denominator. Plot this point.

If m is negative, move *down* the number of spaces in m's numerator and *right* the number of spaces in m's denominator. Plot this point.

Draw the line through the two points plotted.

Example

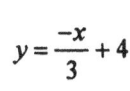
$y = \dfrac{-x}{3} + 4$

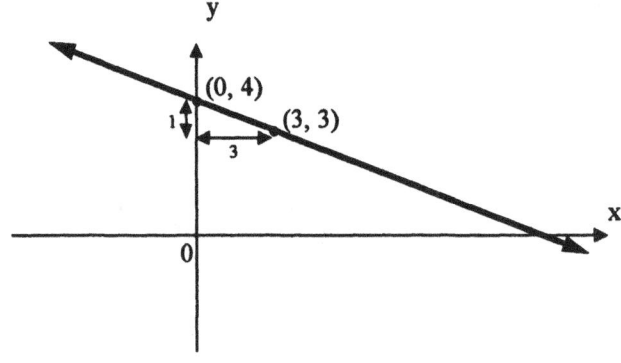

39. Graphing lines of the form $x = a$ and $y = b$

A linear equation having only one variable represents either a vertical line ($x = a$) or a horizontal line ($y = b$). Draw a vertical line through a point $(a, 0)$. Draw a horizontal line through a point $(0, b)$.

Examples

$$x - 4 = 0$$
$$x = 4$$

$$3y = 1$$
$$y = 1/3$$

A Brief Review of Some Essentials of Probability

1. Experiment and outcome
An experiment in probability theory generally refers to any observable activity in which one outcome is equally as likely to occur as another outcome.

Examples

    Experiment: Tossing a fair coin
    Outcome: Heads or tails

    Experiment: Rolling a fair die
    Outcome: The number of dots that turn up

    Experiment: Choosing a playing card at random from a deck of cards
    Outcome: The suit and value of the card

2. Sample space
The set of all possible outcomes in an experiment.

Examples

    The experiment of tossing a coin has sample space {heads, tails}.

    The experiment of rolling a die has sample space {1, 2, 3, 4, 5, 6}.

    The experiment of choosing a playing card has the sample space in which all 52 cards are listed according to face value and suit: {2 of diamonds, 3 of diamonds, ..., 2 of clubs, 3 of clubs, ...}.

3. Event
The particular outcome that the person performing the experiment seeks.

Examples

    The event of a tail turning up when a fair coin is tossed.

    The event of an even number turning up when a fair die is rolled.

    The event of a queen being chosen randomly from a deck of cards.

4. Probability of an event
Given a sample space having n equally likely outcomes and s ways for a particular event to occur, the probability of that event occurring is:

$$p = \frac{s}{n}$$

For any event, since $0 \leq s \leq n$, $0 \leq p \leq 1$.

Examples

The probability of a tail turning up when a fair coin is tossed is ½, or 0.5.

The probability of an even number turning up when a fair die is rolled is 3/6, or 0.5.

The probability of either an even or an odd number turning up when a fair die is rolled is 6/6, or 1. This is a *certain* event.

The probability of a 7 turning up when a fair die is rolled is 0. This is an *impossible* event.

The probability of a queen being randomly chosen from a deck of cards is 4/52, or 1/13.

The probability of a girl being chosen at random from a class of 8 girls and 11 boys is 8/19. (The sample space has 19 equally likely outcomes since there are 19 children.)

5. Counting (multiplication principle)

Suppose there are n_1 *different* ways for an event to occur. Suppose also that there are n_2 ways for a second event to occur. Then the number of ways for the sequence of the first event followed by the second to occur is given by:

$$n_1 \cdot n_2$$

In general, the number of ways for a sequence of events to occur is given by:

$$n_1 \cdot n_2 \cdot \ldots \cdot n_m$$

in which n_1, n_2,... are the number of ways for successive events to occur and m is the number of events in the sequence.

Examples

The number of ways to pick 5 cards from a deck of playing cards (without replacing a card once it is chosen) is:
$$52 \cdot 51 \cdot 50 \cdot 49 \cdot 48 = 311,875,200$$

The number of ways to arrange the letters in the name Alfred is:
$$6 \cdot 5 \cdot 4 \cdot 3 \cdot 2 \cdot 1 = 720$$
(Notice that all the letters are different in this name.)

The number of ways for 10 people to line up is:
$$10! = 10 \cdot 9 \cdot 8 \cdot 7 \cdot \ldots \cdot 2 \cdot 1 = 3,628,800$$

Definition of $n!$ $\qquad n! = n \bullet (n-1) \bullet (n-2) \bullet \cdots \bullet 2 \bullet 1$

A Brief Review of Some Essentials of Geometry

1. Complementary angles (two angles whose measures have a sum of 90^0)

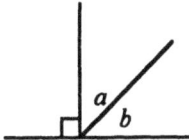

$$m\angle a + m\angle b = 90^0$$

2. Supplementary angles (two angles whose measures have a sum of 180^0)

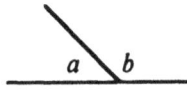

$$m\angle a + m\angle b = 180^0$$

3. An exterior angle of a triangle equals in measure the sum of the measures of the two remote interior angles.

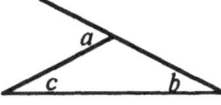

$$m\angle a = m\angle b + m\angle c$$

4. Vertical angles have equal measure.

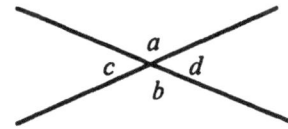

$$m\angle a = m\angle b$$
$$m\angle c = m\angle d$$

5. Perpendicular lines intersect to form right angles.

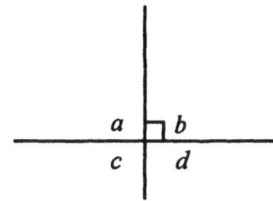

$$m\angle a = m\angle b = m\angle c = m\angle d = 90^0$$

6. An acute angle has measure greater than 0^0 and less than 90^0.

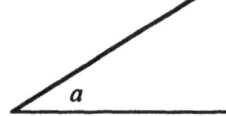

$\angle a$ is acute

7. An obtuse angle has measure greater than $90°$ and less than $180°$.

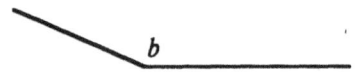

∠b is obtuse

8. Parallel lines (*m* and *n*)

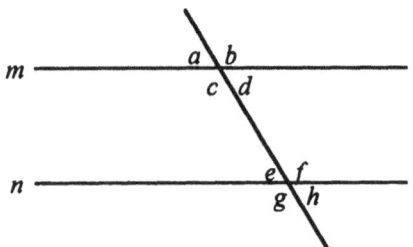

Alternate interior angles have equal measure.
$m\angle c = m\angle f$ and $m\angle d = m\angle e$

Corresponding angles have equal measure.
$m\angle a = m\angle e$, $m\angle b = m\angle f$, $m\angle c = m\angle g$, $m\angle d = m\angle h$

Consecutive interior angles are supplementary.
$m\angle c + m\angle e = 180°$, $m\angle d + m\angle f = 180°$

9. The sum of the measures of the interior angles of any polygon of n sides is $(n - 2)\,180°$.
(For example: The sum of the measures of the angles of a triangle is $(3 - 2)\,180° = 180°$.)

Example

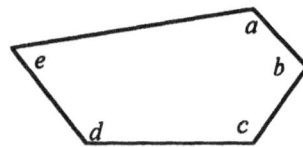

$n = 5$, therefore:
$m\angle a + m\angle b + m\angle c + m\angle d + m\angle e = (5 - 2)\,180° = 540°$

10. The sum of the measures of the exterior angles of any polygon is $360°$.

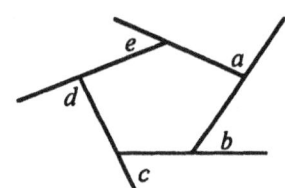

$m\angle a + m\angle b + m\angle c + m\angle d + m\angle e = 360°$

11. The base angles of an isosceles triangle have equal measure (the converse is also true).

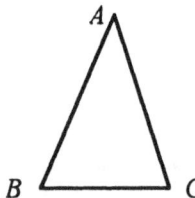

If $AB = AC$, then $m\angle B = m\angle C$.
If $m\angle B = m\angle C$, then $AB = AC$.

12. The Pythagorean theorem:
The sum of the squares of the lengths of the legs of a right triangle equals the square of the length of the hypotenuse.

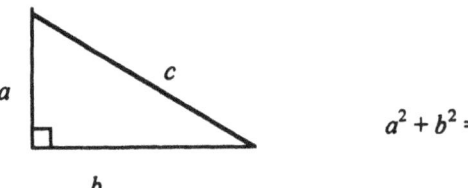

$$a^2 + b^2 = c^2$$

Given any two sides of a right triangle, we can always find the third side by using the Pythagorean theorem. However, problems involving right triangles (especially those on the *SAT I*) almost always use triangles with whole numbers (or whole number ratios) as sides. Therefore, you can save valuable time if you have memorized the more frequently used sets of integers that satisfy the Pythagorean relationship. Then when you spot two numbers from one of these sets, you know the third without doing any arithmetic. The following is a list of the Pythagorean "triples" you should learn to recognize on sight. Remember, too, that multiples of these sets also satisfy the relation.

3-4-5 (6-8-10; $\frac{3}{5}-\frac{4}{5}-\frac{5}{5}$; 9π-12π-15π; etc.)

5-12-13   9-40-41
8-15-17   12-35-37
7-24-25   11-60-61
20-21-29

Example
In solving a problem, you have a right triangle in which one leg has length 10 and the hypotenuse has length 26. Therefore, the other leg has length 24, since the triple 10-24-26 is a multiple of (twice) the Pythagorean triple 5-12-13.

Note
When reduced to their lowest terms, no two members of a Pythagorean triple have a common factor. Therefore, if the two known sides of the triangle have a common factor (as in the example of 10 and 26), reduce by that factor to help identify the Pythagorean triple (if any) to which the sides belong. (Don't forget to multiply the length of the third side in the reduced triple by the same factor.)

13. The area of a triangle is equal to one half the product of the lengths of a base and the altitude drawn to that base.

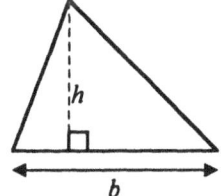 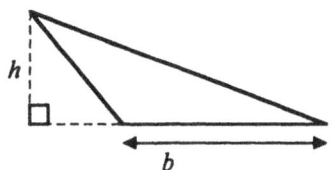

$$\text{area} = \frac{1}{2} bh$$

14. The area of a right triangle is equal to one half the product of the lengths of its legs.

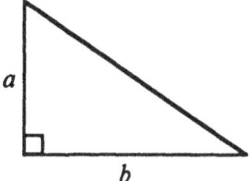

$$\text{area} = \frac{1}{2} ab$$

15. **The area of an equilateral triangle** with side length S is equal to $\frac{S^2\sqrt{3}}{4}$.

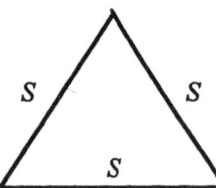

area = $\frac{S^2\sqrt{3}}{4}$

16. **The area of a rectangle** is equal to the product of its length and width.

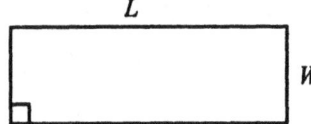

area = LW

17. **The area of a square** is equal to the length of a side squared, OR one-half the length of the diagonal squared.

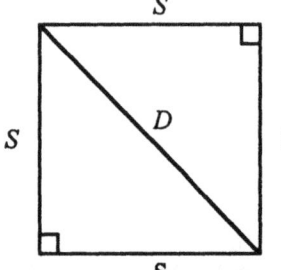

area = S^2
OR
area = $\frac{1}{2}D^2$
Also:
$D = S\sqrt{2}$
and:
$S = \frac{1}{2}D\sqrt{2}$

18. **The area of a parallelogram** is equal to the product of the lengths of its base and altitude drawn to that base.

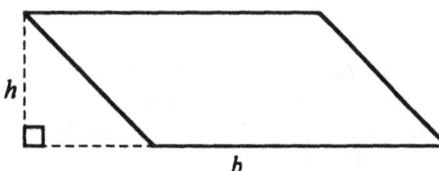

area = bh

19. **The circumference of a circle** with radius r is $2\pi r$.

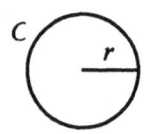

$C = 2\pi r$

20. **The area of a circle** with radius r is πr^2.

area = πr^2

21. Since the **area of a sector** of a circle, radius r, is $\frac{n^0}{360^0}$ of the area of the circle (n is the measure of the central angle of the sector), we get:

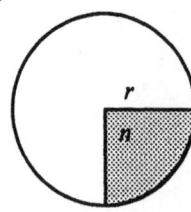

area of sector = $\frac{n^0}{360^0} \cdot \pi r^2$

22. Since the **length of an arc** of a circle, radius r, is $\dfrac{n^0}{360^0}$ of the circumference of the circle (n is the measure of the arc), we get:

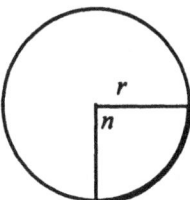

length of arc $= \dfrac{n^0}{360^0} \cdot 2\pi r$

Note:
From Rules 20 and 21 we get the following proportion:

$$\dfrac{n^0}{360^0} = \dfrac{\text{area of sector}}{\text{area of circle}} = \dfrac{\text{length of arc}}{\text{circumference}}$$

23. For **any similar polygons** I and II, we may set up the following proportions:

$$\dfrac{\text{Area I}}{\text{Area II}} = \dfrac{(\text{Side I})^2}{(\text{Corresponding side of II})^2} = \dfrac{(\text{Any linear part I})^2}{(\text{Corresponding linear part II})^2}$$

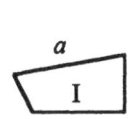 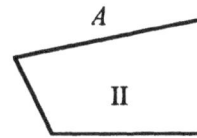

$\dfrac{\text{area I}}{\text{area II}} = \dfrac{a^2}{A^2}$

Similarly for **3-dimensional similar figures**:
Two corresponding volumes are to each other as the cubes of their corresponding edges. That is,

$$\dfrac{\text{Volume I}}{\text{Volume II}} = \dfrac{(\text{Edge I})^3}{(\text{Corresponding edge II})^3}$$

24. Two triangles are similar if (1) the measures of their corresponding angles are equal and (2) the lengths of their corresponding sides are in proportion. If (1) is true, then (2) is also true, and conversely, if (2) is true, then (1) is also true.

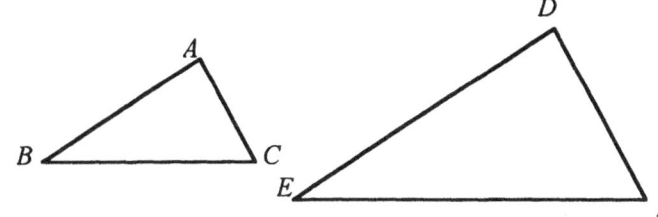

$\triangle ABC$ is similar to $\triangle DEF$
$m\angle A = m\angle D, m\angle B = m\angle E, m\angle C = m\angle F$

$$\dfrac{AB}{DE} = \dfrac{AC}{DF} = \dfrac{BC}{EF}$$

Note:
If a line in a triangle is parallel to one side of the triangle, then it forms a triangle similar to the original triangle.

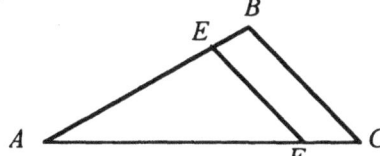

If $\overline{EF}$ is parallel to $\overline{BC}$, $\triangle ABC$ is similar to $\triangle AEF$.

113

25. Circles and angle measurement

A. The measure of a central angle is equal to the measure of its intercepted arc.

$x = a$

B. The measure of an angle with vertex on the circle (i.e., formed by either two chords intersecting on the circle, or a tangent and a chord intersecting on the circle) is equal to one half the measure of its intercepted arc.

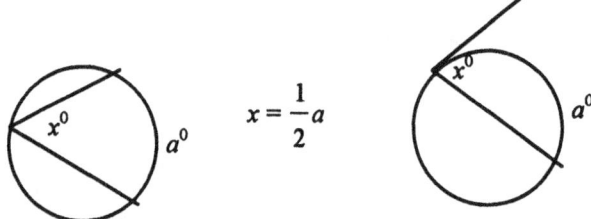

$x = \dfrac{1}{2} a$

C. The measure of an angle formed by two chords intersecting inside the circle equals one half the sum of the measures of the two intercepted arcs.

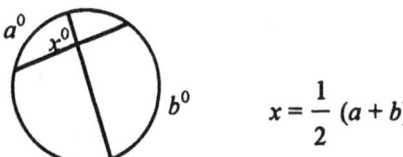

$x = \dfrac{1}{2}(a + b)$

D. The measure of an angle whose vertex is outside the circle is equal to one half the difference of the measures of the two intercepted arcs.

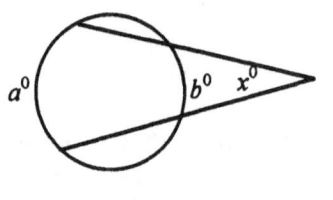 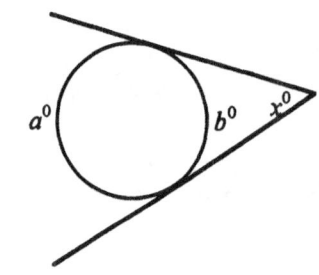

$x = \dfrac{1}{2}(a - b)$

26. Geometric inequalities

A. The measure of an exterior angle of a triangle is greater than the measure of either remote interior angle.

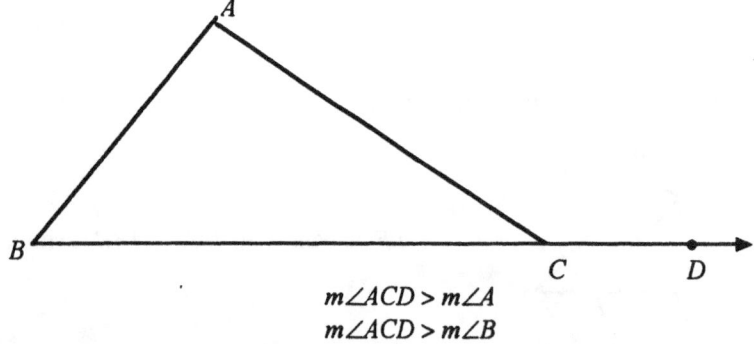

$m\angle ACD > m\angle A$
$m\angle ACD > m\angle B$

B. If two sides of a triangle are not congruent, then the angles opposite those sides are not congruent, the angle with greater measure being opposite the longer side.

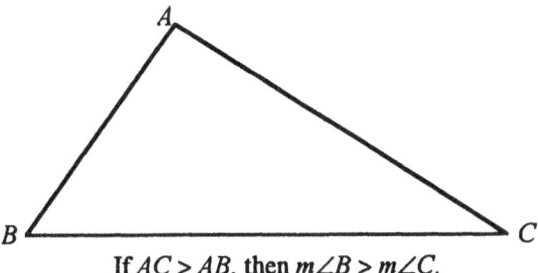

If $AC > AB$, then $m\angle B > m\angle C$.

C. If two angles of a triangle are not congruent, then the sides opposite those angles are not congruent, the longer side being opposite the angle with the greater measure.

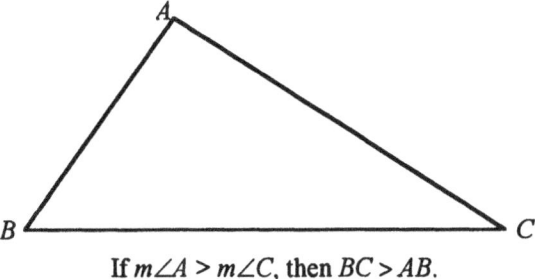

If $m\angle A > m\angle C$, then $BC > AB$.

D. The sum of the lengths of any two sides of a triangle is greater than the length of the third side.

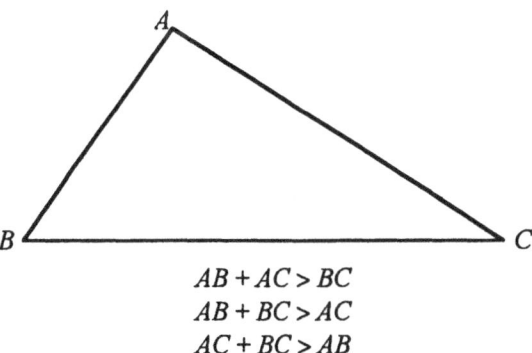

$AB + AC > BC$
$AB + BC > AC$
$AC + BC > AB$

E. The shortest distance between a given point and a given line is the length of the perpendicular segment from the point to the line.

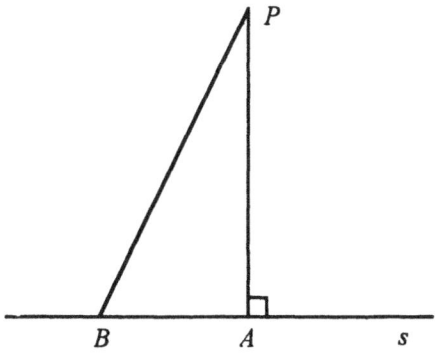

If $\overline{PA}$ is perpendicular to line s, and B is any point on line s other than A, then $PB > PA$.

F. If two sides of a triangle are congruent respectively to two sides of a second triangle, and the measure of the included angle of the first triangle is greater than the measure of the included angle of the second triangle, then the measure of the third side of the first triangle is greater than the measure of the third side of the second triangle.

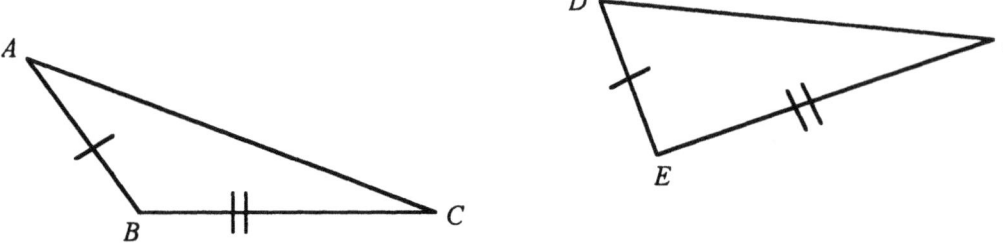

If $AB = DE$ and $BC = EF$, and if $m\angle B > m\angle E$, then $AC > DF$.

G. If two sides of one triangle are congruent respectively to two sides of a second triangle, and the measure of the third side of the first triangle is greater than the measure of the third side of the second triangle, then the measure of the included angle of the first triangle is greater than the measure of the included angle of the second triangle.

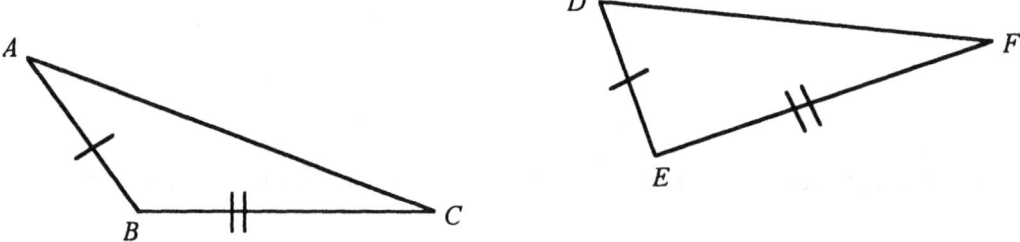

If $AB = DE$ and $BC = EF$, and if $AC > DF$, then $m\angle B > m\angle E$.

The Corwin Press logo—a raven striding across an open book—represents the happy union of courage and learning. We are a professional-level publisher of books and journals for K-12 educators, and we are committed to creating and providing resources that embody these qualities. Corwin's motto is "Success for All Learners."

In compliance with GPSR, should you have any concerns about the safety of this product, please advise: International Associates Auditing & Certification Limited The Black Church, St Mary's Place, Dublin 7, D07 P4AX Ireland EUAR@ie.ia-net.com

www.ingramcontent.com/pod-product-compliance
Lightning Source LLC
Chambersburg PA
CBHW082243300426
44110CB00036B/2426